William Garneys

Contributions to the Flora and Fauna of Repton

and neighbourhood

William Garneys

Contributions to the Flora and Fauna of Repton
and neighbourhood

ISBN/EAN: 9783337273026

Printed in Europe, USA, Canada, Australia, Japan

Cover: Foto ©berggeist007 / pixelio.de

More available books at **www.hansebooks.com**

CONTRIBUTIONS

TO THE

FLORA AND FAUNA

OF

REPTON

AND NEIGHBOURHOOD.

LONDON:
BEMROSE & SONS, 23, OLD BAILEY;
AND IRONGATE, DERBY.

1881.

ALTERUM LATINE.

Quo tendis liber elegantiorum
Florum dotibus aureis venustus?
Quem finxere manu pia sodales
Intextum violæ rosæque sertis,
Ut solatiolum adsit otiosis
Ut lætosque dies, et heu peractæ
Gratam reddat imaginem juventæ.
Quo tu quo, precor, O libelle tendis?
Floræ confugis in sinum? Sapisti:
Illa vindice, neminem timeto.

H. E. F.

PREFACE TO THE FIRST EDITION.

This volume has been compiled for the guidance of the Students of Natural History in Repton School, in the hope that it may receive many additions and corrections as further researches are carried on.

For the writers it contains a record of pleasant hours, passed amidst scenes known and dear to all old Reptonians. Few Schools in England enjoy such advantages as we possess in the beauty of our neighbourhood and the extent of our range. The very names of Bretby and Foremark, Knoll Hills and Anchor Church, Repton Rocks, Twyford Ferry, Tanyard Lane, call up a landscape of rich and varied interest, noble timber and fragrant lanes, broad meadows

"dewy fresh, browsed by deep-uddered kine,"

and woods carpeted with bracken, or glowing with blue-bells; and the "smug and silver Trent," always fresh and teeming with life, winding like a bright riband through the wide valley.

To all Repton men these scenes are connected with pleasant recollections, but surely to none so much as those whose enjoyment of liberty and exercise has been enhanced by a love of Nature, rendered distinct and permanent by a definite pursuit. "Happy truly is the Naturalist. The earth becomes to him transparent; everywhere he sees significances, harmonies, laws, chains of cause and effect endlessly interlinked, which draw him out of the narrow sphere of self-interest and self-pleasing into a pure and wholesome region of solemn joy and wonder."*

By some such these pages have been arranged, and by all will be received with a kindly welcome, and meet with gentle criticism. Let us hope that this little volume may be the companion of many a pleasant ramble over the old ground. Let it also convey a caution to the hunter of plants or insects, eggs or shells, that in his own keen pursuit he may recollect to leave something for those who come after, and carefully to preserve all rare specimens. He may at the same time be reminded, while himself enjoying abundant liberty of movement, to regard the claims of property, and not to forget the rights of his neighbours.

The Flora is the work of W. Wyatt and C. G. Thornton, from notes of Collections made in 1859, 1860, and 1861, with additions by W. M. Sinclair, E. Blumhardt, and W. Nanson. Much assistance has been given by A. Hewgill, Esq., M.D., to whose exact and extensive knowledge, and constant encouragement, the study of Botany in Repton School owes its introduction and successful prosecution.

The List of Birds has been contributed by A. O. Worthington, whose early propensity for birds-nesting has developed into a wide and accurate knowledge of Ornithology.

The section of Moths and Butterflies is the work of the Rev. F. Spilsbury, whose attainments as an entomologist are too well known to need any commendation from the writer of these introductory remarks.

* Rev. C. Kingsley.

PREFACE TO THE SECOND EDITION.

"In 1873 the first edition had been exhausted, and it became necessary to take steps to publish a second. Nothing definite, however, was done until the next year, when F. O. Bower made an enlarged List of Plants. This List was in 1875 revised by Mr. Garneys, and considerable additions made by that gentleman, and finally the corrected List compiled by H. Lewis.

"In the first edition the classification of the Rev. C. A. John in "Flowers of the Field" had been adopted. Since the publication of that book the views of Botanists on many important points have changed, and the present edition of the Repton Flora is based on the Handbook of British Flora, by G. Bentham, F.R.S., published in 1866."

Thus wrote Mr. Lewis in 1875, and the second edition of the "Repton Flora" has not been published until now. I cannot tell who is responsible for the delay, nor is it necessary, I think, to enter into the reason thereof; sufficient is it to say that I have once more carefully revised the "List of Plants," with Mr. Hagger's assistance, and we have yet again added some few more species. Mr. A. O. Worthington has revised his "List of Birds," and, in his own words, trusts that he has made it worthy of the "Old School." I have, with the help of Mr. Mason, of Burton-on-Trent, fairly increased the "List of Lepidoptera."

With regard to the additional "Lists" of this volume, those of the "Mosses" and "Mollusca" are equally the work of Mr. Hagger and myself; for that of the "Coleoptera" I am alone responsible. I gratefully acknowledge the assistance of Mr. Hagger in helping me to revise the proofs of this work for the press.

I trust that the publication of these "Contributions" may act as a stimulus to the prosecution of the delightful study of Natural History in Repton School. I can confidently promise that many discoveries are yet to be made, and I can also promise that every possible help and assistance will be rendered by Mr. Hagger and myself.

In conclusion, I have only to say that if any one, on finding a new species to the Repton District, will kindly report to me, I shall be very greatly obliged.

WILLIAM GARNEYS.

Repton, April, 1881.

LIST OF THE PLANTS

FOUND

AT AND NEAR REPTON.

CLASS I.—DICOTYLEDONES.

SUB-CLASS I.—THALAMIFLORÆ.

ORDER I.—RANUNCULACEÆ.

I.—CLEMATIS.

1. *C. vitalba.* Traveller's Joy. 6-8. Ticknall Quarry.

II—THALICTRUM.

3. *T. flavum.* Yellow Meadow Rue. 6-7. Green Lane and Bathing-field Lane.

III.—ANEMONE.

1. *A. nemorosa.* Wood Anemone. 3-5. Repton Shrubs.

V.—MYOSURUS.

1. *M. minimus.* Common Mousetail. 5. Gravelly soil.

VI.—RANUNCULUS.

1. *R. aquatilis.* Water Ranunculus. 5-7.
Old and New Trent.

2. *R. hederaceus.* Ivy-leaved Ranunculus. 5-7.
Brook by Cricket-field Wall.

4. *R. Flammula.* Spearwort. 5-8.
Brook running through the Village.

6. *R. Ficaria.* Lesser Celandine. 3-5.
Everywhere.

7. *R. sceleratus.* Celery-leaved Ranunculus. 6-8.
Brook-end.

8. *R. auricomus.* Goldilocks. Spring.
Bathing-field Lane.

9. *R. acris.* Meadow Ranunculus. 5-7.
Fields.

10. *R. repens.* Creeping Ranunculus. 5-8.
Ditches by Old Trent.

11. *R. bulbosus.* Bulbous Ranunculus. 5-6.
Meadows.

*12. *R. philonotis.* Hairy Ranunculus. 5-6.
Field on the way to Anchor Church.

13. *R. parviflorus.* Small-flowered Ranunculus. 5-8.
Gravelly soil.

14. *R. arvensis.* Corn Ranunculus. 6.
Corn-fields.

VII.—CALTHA.

1. *C. palustris.* Marsh Marigold. 4-6.
Moist Meadows.

---

* This is *R. hirsutus* of London Catalogue.

ORDER III.—NYMPHÆACEÆ.

I.—NYMPHÆA.

1. *N. alba.* White Water Lily. 7. Calko Park.

II.—NUPHAR.

1. *N. lutea.* Yellow Water Lily. Old Trent.

ORDER IV.—PAPAVERACEÆ.

I.—PAPAVER.

2. *P. Rhœas.* Field Poppy. 6-7. Corn-fields.
3. *P. dubium.* Long-headed Poppy. 6-7. Churchyard Wall.
5. *P. Argemone.* Pale Poppy. 5-7. Corn-fields.

II.—CHELIDONIUM.

C. majus. Common Celandine. 5-8. Repton Village.

ORDER V.—FUMARIACEÆ.

I.—FUMARIA.

1. *F. officinalis.* Common Fumitory. 5-9. Corn-fields.

ORDER VI.—CRUCIFERÆ.

I.—MATTHIOLA.

1. *M. incana.* Common Stock. 6-8.
Egginton Road—from refuse from a garden.

II.—CHEIRANTHUS.

1. *C. Cheiri.* Wallflower. 5-6.
The Priory Garden Wall.

III.—BARBAREA.

1. *B. vulgaris.* Common Wintercress. 6-8.
Brook.

IV.—NASTURTIUM.

1. *N. officinale.* Common Watercress. 6-8.
Old Trent.
2. *N. sylvestre.* Creeping Watercress. 6-8.
Banks of New Trent.
3. *N. palustre.* Marsh Watercress. 6-9.
Twyford Ferry.
4. *N. amphibium.* Great Watercress. 5-8.
Upper Bathing-place.

V.—ARABIS.

1. *A. perfoliata.* Glabrous Rockcress. 5-6.
Roadsides.
5. *A. thaliana.* Thale Rockcress. 3-7.
Cricket-field Wall.

VI.—CARDAMINE.

1. *C. amara.* Large Bittercress. 4-5.
Osier-beds by Old Trent.

2. *C. pratensis.* Cuckoo-flower. 5.
Low Meadows.

3. *C. hirsuta.* Hairy Bittercress. 5-8.
Brook under Cricket-field Wall.

VIII.—SISYMBRIUM.

1. *S. officinale.* Common Hedge Mustard. 6-7.
Tanyard Lane.

IX.—ALLIARIA.

1. *A. officinalis.* Garlic Mustard. 4-6.
Field between Bathing-field and Bathing-field Lane.

XI.—BRASSICA.

2. *B. muralis.* Sand Brassica. 5-7.
Goods Station near Willington Junction.

5. *B. campestris.* Field Brassica. 6-7.
Askew Hill.

6. *B. alba.* Mustard Brassica. 6.
Waste Grounds.

7. *B. Sinapistrum.* Charlock. 5-8.
Farmyard facing Cricket-field.

8. *B. nigra.* Black Brassica. 6-7.
Waste Grounds.

XIV.—DRABA.

5. *D. verna.* Whitlow Grass. 2-5.
Wall above Crewe's Pond.

XXI.—CAPSELLA.

1. *C. Bursa-pastoris.* Shepherd's Purse Capsel. 4-9.
Upper Paddock.

XXII.—LEPIDIUM.

1. *L. campestre.* Field Cress. 5-7. Cultivated Fields.
3. *L. ruderale.* Narrow-leaved Cress. 5-7. Near Willington Junction.

XXIII.—SENEBREIA.

1. *S. Coronopus.* Wartcress. 7-9. Waste Ground.

XXVII.—RAPHANUS.

1. *R. Raphanistrum.* Radish. 5-7. Fields near Burton Road.

ORDER VII.—RESEDACEÆ.

I.—RESEDA.

1. *R. Luteola.* Dyer's Rocket. 7. Ticknall Quarry.

ORDER IX.—VIOLACEÆ.

1.—VIOLA.

1. *V. palustris.* Marsh Violet. 4-6. Marsh at Repton Rocks.
2. *V. odorata.* Sweet Violet. 4-5. Burton and Milton Roads.
3. *V. canina.* Dog Violet. 4-7. Banks.
4. *V. tricolor.* Pansy Violet. 5-8. Corn-fields.

ORDER X.—POLYGALACEÆ.

I.—POLYGALA.

1. *P. vulgaris.* Common Milkwort. 5-8. Repton Rocks.

ORDER XII.—CARYOPHYLLACEÆ.

III.—SILENE.

2. *S. inflata.* Bladder Campion. 6-8. Gravel-pit in Repton Village.

7. *S. noctiflora.* Night-flowering Catchfly. 7. Repton Rocks.

IV.—LYCHNIS.

1. *L. vespertina.* White Lychnis. 5-8. Gravel-pit in Repton Village.

2. *L. diurna.* Red Lychnis. 5-8. Roadsides.

3. *L. Githago.* Corn Cockle. 6-7. Corn-fields.

4. *L. Flos-cuculi.* Ragged Robin. 4-7. Osier-beds.

V.—SAGINA.

1. *S. procumbens.* Procumbent Pearlwort. 5-8. Walls and Banks.

VII.—ARENARIA.

5. *A. serpyllifolia.* Thyme-leaved Sandwort. 6-8. School-walls.

7. *A. trinervis.* Three-nerved Sandwort. 5-7. Gravel-pit in Repton Village.

VIII.—MŒNCHIA.

1. *M. erecta.* Upright Mœnchia. 5.
Gravelly soil.

X.—CERASTIUM.

1. *C. vulgatum.* Mouse-ear Chickweed. 4-9.
Everywhere.

2. *C. arvense.* Field Mouse-ear Chickweed. 6-7.
Fields.

XI.—STELLARIA.

1. *S. aquatica.* Water Starwort. 7.
By the old Trent.

3. *S. media.* Chickweed Starwort. 4-7.
Fields and Roadsides.

4. *S. uliginosa.* Bog Starwort. 6.
Marshy places.

5. *S. graminea.* Lesser Stitchwort. 5-7.
Hedges.

7. *S. Holostea.* Stitchwort. 4-6.
Hedges.

XII.—SPERGULARIA.

1. *S. rubra.* Common Sandspurry. 7-8.
Sandy places.

XIII.—SPERGULA.

1. *S. arvensis.* Corn Spurry. 5-8.
Corn-fields.

ORDER XIII.—PORTULACEÆ.

II.—MONTIA.

1. *M. fontana.* Blinks. 4-7.
Repton Rocks.

ORDER XVI.—HYPERICINEÆ.

I.—HYPERICUM.

1. *H. calycinum.* Large flowered Hypericum. 5-8.
Knoll Hills.

3. *H. perforatum.* St. John's wort. 7-8.
Tanyard Lane.

5. *H. quadrangulum.* Square-stalked Hypericum. 7-8.
Ditches under Parson's Hills.

6. *H. humifusum.* Trailing Hypericum. 6-8.
Egginton Road.

8. *H. pulchrum.* Slender Hypericum. 7-8.
Field between Milton and Repton Rocks.

9. *H. hirsutum.* Hairy Hypericum. 7-8.
Repton.

10. *H. montanum.* Mountain Hypericum. 6-8.
Ticknall Quarry.

ORDER XVII.—LINACEÆ.

I.—LINUM.

1. *L. usitatissimum.* Common Flax. 6-7.
Lane down the Village.

3. *L. angustifolium.* Pale Flax. 7-9.
Between Willington and Etwall.

4. *L. catharticum.* Cathartic Flax. 5-8.
Dry pastures.

ORDER XVIII.—MALVACEÆ.

II.—MALVA.

1. *M. rotundifolia.* Dwarf Mallow. 6-9.
   Under the Cricket-field wall in the Upper Paddock.
2. *M. sylvestris.* Common Mallow. 6-8.
   Roadsides.
3. *M. moschata.* Musk Mallow. 7-8.
   Road to Ticknall.

ORDER XIX.—TILIACEÆ.

I.—TILIA.

1. *T. Europœa.* Common Lime. 7.
   Upper Paddock.

ORDER XX.—GERANIACEÆ.

I.—GERANIUM.

4. *G. pratense.* Meadow Geranium. 6-8.
   Osier-beds.
6. *G. Robertianum.* Herb Robert Geranium. 5-8.
   Tanyard Lane, etc.
7. *G. lucidum.* Shining Geranium. 5-8.
   Cricket-field Wall.
8. *G. molle.* Dove's-foot Geranium 5-8.
   Upper Paddock.
9. *G. pusillum.* Small-flowered Geranium. 5-8.
   Upper Paddock.

10. *G. rotundifolium.* Round-leaved Geranium. 6-7.
Burton Road.

11. *G. dissectum.* Cut-leaved Geranium. 5-8.
Lane at the top of the Hall Orchard.

12. *G. columbinum.* Long-stalked Geranium. 5-8.
Burton Road.

II.—ERODIUM.

1. *E. cicutarium.* Common Erodium. 5-8.
Road between Milton and Repton Rocks.

III.—OXALIS.

1. *O. Acetosella.* Wood Sorrel. 5-6.
Woods.

ORDER XXI.—ACERACEÆ.

I.—ACER.

1. *A. campestre.* Common Maple. 5-6.
Bathing-field Lane.

2. *A. Pseudo-platanus.* Sycamore. 5-6.
Foremark.

SUB-CLASS II.—CALYCIFLORÆ.

ORDER XXII.—AQUIFOLIACEÆ.

I.—ILEX.

1. *I. aquifolium.* Common Holly. 5-6.
Milton Road.

ORDER XXIII.—CELASTRACEÆ.

I.—EUONYMUS.

1. *E. europœus.* Common Spindle-tree. 5. Repton Rocks.

ORDER XXIV.—RHAMNACEÆ.

I.—RHAMNUS.

2. *R. Frangula.* Alder Buckthorn. 5. Repton Rocks.

ORDER XXV.—PAPILIONACEÆ.

I.—ULEX.

1. *U. europœus.* Common Furze. 2-6. Repton Rocks.
2. *U. Gallii.* Late Northern Furze. 9-10. Repton Rocks.

II.—GENISTA.

1. *G. tinctoria.* Dyers' Green-weed. 7-8. (A. H.)
3. *G. anglica.* Needle Genista. 4-5. Etwall Road.

III.—SAROTHAMNUS.

1. *S. scoparius.* Common Broom. 6. Repton Rocks.

IV.—ONONIS.

1. *O. arvensis.* Restharrow. 5-8. Field beyond the Stein-yard.

V.—MEDICAGO.

3. *M. lupulina.* Black Medick. 6-8.
Askew Hill.

VI.—MELILOTUS.

1. *M. officinalis.* Common Melilot. 6-7.
Lane at the top of the Hall Orchard.

VIII.—TRIFOLIUM.

1. *T. arvense.* Hare's-foot Clover. 7-9.
Gravel-pit in Repton Village.

4. *T. ochroleucum.* Sulphur Clover. 7-8.
Dry pastures.

5. *T. pratense.* Purple Clover. 5-8.
Fields.

6. *T. medium.* Zig-zag Clover. 7-8.
Dry gravelly pastures.

8. *T. striatum.* Knotted Clover. 6-7.
Dry pastures.

10. *T. scabrum.* Rough Clover. 5-8.
(W. G.)

15. *T. subterraneum.* Subterranean Clover. 6-7.
Dry pastures.

17. *T. repens.* White Clover. 5-8.
Gravel-pit in Repton Village.

19. *T. procumbens.* Hop Clover. 6-7.
Gravel-pit in Repton Village.

21. *T. filiforme.* Slender Clover. 6-7.
Field near Milton.

IX.—LOTUS.

1. *L. corniculatus.* Bird's-foot Trefoil. 7-8.
Osier-beds and Gravel-pit in Repton Village.

X.—ANTHYLLIS.

1. *A. vulneraria.* Kidney Vetch. 6-7.
Willington Road.

XIII.—ORNITHOPUS.

2. *O. perpusillus.* Common Bird's-foot. 6-8.
Gravel-pit in Repton Village.

XVI.—VICIA.

1. *V. hirsuta.* Hairy Vetch. 6-7.
Way to Twyford.

2. *V. tetrasperma.* Slender Vetch. 6-7.
On Canal Bank between Willington Station and Junction.

3. *V. Cracca.* Tufted Vetch. 7-8.
Bathing-field Lane.

4. *V. sylvatica.* Wood Vetch. 7-8.
Repton Rocks.

6. *V. sepium.* Bush Vetch. 5-6.
Banks and Roadsides.

8. *V. sativa.* Common Vetch. 6-7.
Burton Road.

9. *V. lathyroides.* Spring Vetch. 4-5.
(A. H.)

XVII.—LATHYRUS.

4. *L. pratensis.* Meadow Pea. 6-8.
Osier-beds by Willington Road.

9. *L. macrorrhizus.* Tuberous Pea. 6-7.
Repton Rocks.

ORDER XXVI.—ROSACEÆ.

I.—PRUNUS.

1. *P. communis.* Blackthorn. 3-4.
Green Lane.

2. *P. Cerasus.* Cherry. 5.
Bank of Churchyard facing Old Trent.

3. *P. Padus.* Birdcherry. 5.
Copse on left of Burton Road.

II.—SPIRÆA.

2. *S. Ulmaria.* Meadow-sweet. 7-8.
Osier-bed.

3. *S. Filipendula.* Dropwort. 7-9.
Four fields from the Bathing-field towards Parson's Hills.

IV.—GEUM.

1. *G. urbanum.* Herb Bennet. 6-8.
Burton Road, &c.

V.—RUBUS.

1. *R. idæus.* Raspberry. 5-6.
Road to Repton Rocks.

2. *R. fruticosus.* Blackberry. 6-8.
Tanyard Lane, etc.

3. *R. cæsius.* Dewsbury. 5-6.
Road from Repton Rocks to Milton.

VI.—FRAGARIA.

1. *F. vesca.* Strawberry. 6-7.
Cricket-field Wall, etc.

VII.—POTENTILLA.

1. *P. Fragariastrum.* Strawberry-leaved Potentil. 1-6.
Parson's Hills.

2. *P. reptans.* Cinquefoil. 6-8.
Willington Road.

3. *P. Tormentilla.* Tormentil Potentil. 5-8.
Repton Rocks.

7. *P. anserina.* Silver-weed. 6-7.
Brook-end.

9. *P. comarum.* Marsh Potentil. 7.
Repton Rocks.

IX.—ALCHEMILLA.

1. *A. vulgaris.* Lady's-mantle. 6-8.
Osier-beds.

3. *A. arvensis.* Parsley Piert. 5-8.
Corn-fields on Parson's Hills.

X.—SANGUISORBA.

1. *S. officinalis.* Great Burnet. 5-9.
Osier-beds.

XI.—POTERIUM.

1. *P. sanguisorba.* Salad Burnet. 7-8.
Field near Willington Bridge.

XII.—AGRIMONIA.

1. *A. Eupatoria.* Common Agrimony. 7-8.
Milton Road.

XIII.—ROSA.

2. *R. villosa.* Downy Rose. 6-7.
(A. H.)

3. *R. rubiginosa.* Sweetbriar Rose. 6-7.
(A. H.)

4. *R. canina.* Dog Rose. 6-7.
Bathing-field Lane.

5. *R. arvensis.* Field Rose. 6-8.
Bathing-field Lane.

XIV.—PYRUS.

1. *P. communis.* Pear-tree. 4-5.
Anchor Church.

2. *P. Malus.* Crab-tree. 5.
Milton Road.

3. *P. Aria.* White Beam-tree. 5-6.
Foremark.

5. *P. Aucuparia.* Mountain Ash. 4-6.
Repton Rocks.

XV.—CRATÆGUS.

1. *C. Oxycantha.* Hawthorn. 5.
Milton Road.

ORDER XXVII.—ONAGRACEÆ.

I.—EPILOBIUM.

1. *E. angustifolium.* Rose-bay. 7-8.
Egginton Road.

2. *E. hirsutum.* Great Willow-herb. 7-8.
Osier-beds.

3. *E. parviflorum.* Hoary Epilobe. 7-8.
Road to Ticknall.

4. *F. montanum.* Broad Epilobe. 7-8.
Entrance to Gravel-pit in Repton.

5. *E. roseum.* Pale Epilobe. 7-8.
Boggy ground.

6. *E. tetragonium.* Square Epilobe. 7-8.
Osier-bed at Foremark Hall.

7. *E. palustre.* Marsh Epilobe. 7-8.
Marshy ground at Gravel-pit in Repton.

IV.—CIRCÆA.

1. *C. lutetiana.* Enchanter's Nightshade. 7-8.
Between the Cricket-field Wall and Brook.

ORDER XXVIII.—LYTHRARIEÆ.

I.—LYTHRUM.

1. *L. Salicaria.* Purple Loosestrife. 7-8.
Old Trent by Culvert.

II.—PEPLIS.

1. *P. Portula.* Water Purslane. 7-8.
Watery Places.

ORDER XXIX.—CUCURBITACEÆ.

I.—BRYONIA.

1. *B. dioica.* Common Bryony. 5-6.
Hedge of Tanyard.

ORDER XXX.—CRASSULACEÆ.

II.—COTYLEDON.

1. *C. Umbilicus.* Pennywort. 6-8.
   Anchor Church.

III—SEDUM.

5. *S. album.* White Sedum. 7-8.
   Wall in Milton Village.

7. *S. acre.* Biting Sedum. 6-7.
   Wall facing Cricket-field.

9. *S. rupestre.* Rock Sedum. 7-8.
   Cricket field Wall.

IV.—SEMPERVIVUM.

1. *S. tectorum.* Common Houseleek. 7.
   Farm-buildings facing Cricket-field.

ORDER XXXI.—RIBESIACEÆ.

I.—RIBES.

1. *R. Grossularia.* Gooseberry. 4-5.
   Hedge by the Culvert.

2. *R. rubrum.* Red Currant. 4-5.
   Osier-bed through which footpath to Newton passes.

4. *R. nigrum.* Black Currant. 4-5.
   Hedges near Stein-yard.

ORDER XXXII.—SAXIFRAGACEÆ.

I.—SAXIFRAGA.

6. *S. granulata.* Meadow Saxifrage. 5-6.
   Parson's Hills.

9. *S. tridactylites.* Rue-leaved Saxifrage. 5-6. Cricket-field Wall.

II.—CHRYSOSPLENIUM.

1. *C. oppositifolium.* Golden Saxifrage. 4-5. Rocks below the Churchyard.

2. *C. alternifolium.* Alternate Chrysosplene. 5.

III.—PARNASSIA.

1. *P. palustris.* Grass-of-Parnassus. 7-9. Milton.

ORDER XXXIII.—DROSERACEÆ.

I.—DROSERA.

1. *D. rotundifolia.* Common Sundew. 7-8. Repton Rocks.

ORDER XXXIV.—HALORAGEÆ.

I.—MYRIOPHYLLUM.

2. *M. verticillatum.* Whirled Water Milfoil. 7. Ponds.

II.—HIPPURIS.

1. *H. vulgaris.* Common Marestail. 6-7. Swarkestone Bridge.

ORDER XXXV.—UMBELLIFERÆ.

I.—HYDROCOTYLE.

1. *H. vulgaris.* Marsh Pennywort. 6-7. Repton Rocks.

II.—SANICULA.

1. *S. europæa.* Wood Sanicle. 5-7.
Repton Shrubs.

V.—CICUTA.

1. *C. virosa.* Cowbane. 7-8.
Ditches.

VII.—HELOSCIADIUM.

1. *H. nodiflorum.* Procumbent Helosciad. 7-8.
Muddy ditches.

2. *H. inundatum.* Lesser Helosciad. 7-8.
Old Trent.

XI.—ÆGOPODIUM.

1. *Æ. Podagraria.* Goutweed. 5-6.
Milton and Findern Streets.

XII.—CARUM.

2. *C. Carui.* Caraway. 6.
Gravel-pit down the Village.

XIII.—SIUM.

1. *S. latifolium.* Water Parsnip. 6-8.
(A. H.)

2. *S. angustifolium.* Lesser Sium. 7-8.
Marsh near Willington Junction.

XIV.—PIMPINELLA.

1. *P. saxifraga.* Burnet Saxifrage. 7-8.
Parson's Hills.

XVI.—ŒNANTHE.

1. *Œ. fistulosa.* Water Dropwort. 7-8.
Ditch at Bottom of Parson's Hills.

3. *Œ. crocata.* Hemlock Œnanth. 6-8. (A. H.)

4. *Œ. Phellandrium.* Fine-leaved Œnanth. 6-7. Swamp near Willington Bridge.

XVII.—ÆTHUSA.

1. *Æ. Cynapium.* Fool's Parsley. 7-8. Near Anchor Rocks.

XXI.—SILAUS.

1. *S. pratensis.* Pepper Saxifrage. 8-9. Moist meadows.

XXIV.—ANGELICA.

1. *A. sylvestris.* Wild Angelica. 7. Field under Anchor Wood, etc.

XXV.—PEUCEDANUM.

2. *P. palustre.* Milk Parsley. 6-7. Marshes.

XXVI.—PASTINACA.

1. *P. sativa.* Common Parsnip. 7-8. Calke Park.

XXVII.—HERACLEUM.

1. *H. Sphondylium.* Cow Parsnip. 7. Meadows.

XXIX.—SCANDIX.

1. *S. Pecten.* Shepherd's Needle. 6-9. Corn-fields.

XXXI.—BUNIUM.

1. *B. flexuosum.* Tuberous Bunium. 5-6. Hall Orchard.

FLOWERING PLANTS.

XXXII.—CHÆROPHYLLUM.

1. *C. temulum.* Rough Chervil. 6-7. Gravel-pit in Village.
2. *C. sylvestre.* Wild Chervil. 4-6. Mount Pleasant.
3. *C. Anthriscus.* Bear Chervil. 5. Waste ground in the Village.

XXXIII.—CAUCALIS.

1. *C. nodosa.* Knotted Caucalis. 5-7. Way-sides.
2. *C. Anthriscus.* Hedge Parsley. 6-8. Hedges.
3. *C. infesta.* Spreading Caucalis. 7-8. Way-sides.
4. *C. daucoides.* Small Caucalis. 6-9. Milton Road.

XXXIV.—DAUCUS.

1. *D. Carota.* Common Carrot. 7-8. Gravel-pit down the Village.

XXXV.—CONIUM.

1. *C. maculatum.* Common Hemlock. 6-7. Ticknall Quarry and Fox-covers at Findern.

ORDER XXXVI.—ARALIACEÆ.

I.—HEDERA.

1. *H. Helix.* Common Ivy. 10-11. Priory and big Schoolroom.

ORDER XXXVIII.—CORNACEÆ.

I.—CORNUS.

2. *C. sanguinea.* Dogwood. 6.
Osier-bed, through which footpath to Newton passes.

SUB-CLASS III.—MONOPETALOIDEÆ.

ORDER XXXIX.—CAPRIFOLIACEÆ.

I.—ADOXA.

1. *A. Moschatellina.* Tuberous Moscatel. 4-5.
Field below Hall Orchard.

II.—SAMBUCUS.

1. *S. nigra.* Common Elder. 6.
Tanyard Lane.

III.—VIBURNUM.

1. *V. Lantana.* Wayfaring-tree. 5-6.
Repton Shrubs.

2. *V. Opulus.* Guelder Rose. 6-7.
Repton Rocks.

IX.—LONICERA.

1. *L. Periclymenum.* Woodbine. 7.
Milton Road.

ORDER XL.—STELLATÆ.

II.—GALIUM.

1. *G. cruciatum.* Cross-wort. 5-6.
Willington Road.

FLOWERING PLANTS.

2. *G. verum.* Yellow Galium. 7-8. Cricket-field Wall.
3. *G. palustre.* Marsh Galium. 7-8. Field below the Tanyard.
4. *G. uliginosum.* Swamp Galium. 7-8. Near Old Trent.
5. *G. saxatile.* Heath Galium. 7-8. Repton Rocks.
6. *G. Mollugo.* Hedge Galium. 7-8. Hedges.
7. *G. parisiense.* Wall Galium. 6-9. (W. G.)
9. *G. Aparine.* Goosegrass. 6-8. Hedges.

III.—ASPERULA.

1. *A. odorata.* Sweet Woodruff. 5. Repton Shrubs.

IV.—SHERARDIA.

1. *S. arvensis.* Field Madder. 6-8. Fields round Repton Rocks.

ORDER XLI.—VALERIANEÆ.

II.—VALERIANA.

1. *V. dioica.* Marsh Valerian. 6. Repton Shrubs.
2. *V. officinalis.* Common Valerian. 6-7. Tanyard Osier-bed.

III.—VALERIANELLA.

1. *V. olitoria.* Common Cornsalad. 6-7.
Cornfield between Repton Shrubs and Repton Rocks.

4. *V. dentata.* Narrow-leaved Cornsalad. 6-7.
Cornfield between Repton Shrubs and Repton Rocks.

ORDER XLII.—DIPSACEÆ.

I.—DIPSACUS.

1. *D. sylvestris.* Common Teasel. 6-7.
Ticknall Quarry.

2. *D. pilosus.* Small Teasel. 7-9.
Repton Shrubs, near Bretby Mill.

II.—SCABIOSA.

1. *S. succisa.* Blue Scabious. 7.
Repton Shrubs.

2. *S. columbaria.* Small Scabious. 6-8.
Goods Station near Willington Junction.

3. *S. arvensis.* Field Scabious. 7-8.
Field between Repton Shrubs and Repton Rocks.

ORDER XLIII.—COMPOSITÆ.

I.—EUPATORIUM.

1. *E. cannabinum.* Hemp Agrimony. 7-8.
Brook between Milton and Foremark.

II.—TUSSILAGO.

1. *T. Farfara.* Coltsfoot. 3-4.
Ploughed fields on Askew Hill.

2. *T. Petasites.* Butterbur. 4-5.
Osier-bed by Tanyard.

IV.—ERIGERON.

1. *E. acris.* Common Erigeron. 6-9.
Ticknall Quarry.

VI.—SOLIDAGO.

1. *S. Virga-aurea.* Goldenrod. 7-9.
Dry woods.

VII.—INULA.

1. *I. dysenterica.* Fleabane. 7-8.
Marshy ground near Repton Mill.

VIII.—BELLIS.

1. *B. perennis.* Common Daisy. 2-11.
School-yard.

IX.—CHRYSANTHEMUM.

1. *C. Leucanthemum.* Ox-eye Daisy. 5-7.
Meadows by the Old Trent.

2. *C. segetum.* Corn Marigold. 6-9.
Corn-fields, especially near Egginton.

3. *C. Parthenium.* Feverfew Chrysanthemum. 7-8.
Anchor Church. Road to Bretby.

4. *C. inodorum.* Scentless Chrysanthemum. 7-10.
Burton Road.

X.—MATRICARIA.

1. *M. Chamomilla.* Wild Camomile. 6-8.
Roadsides.

XI.—ANTHEMIS.

2. *A. arvensis.* Corn Camomile. 6-7.
Cultivated fields.

XII.—ACHILLEA.

1. *A Ptarmica.* Sneezewort. 7-8.
Egginton Road.

2. *A. Millefolium.* Milfoil. 6-9.
Cricket-field Wall.

XIV.—TANACETUM.

1. *T. vulgare.* Tansy. 7-8.
Anchor Rocks.

XV.—ARTEMISIA.

3. *A. vulgaris.* Mugwort. 7-9.
Bank between Milton and Anchor Church.

XVI.—GNAPHALIUM.

4. *G. sylvaticum.* Wood Cudweed. 7.
Gravelly Pastures.

6. *G. uliginosum.* Marsh Cudweed. 8-9.
Near Repton Rocks.

7. *G. germanicum.* Common Cudweed. 6-7.
Field between Milton and Repton Rocks.

8. *G. arvense.* Field Cudweed, 7-8.
Field near Repton Rocks.

XVII.—SENECIO.

1. *S. vulgaris.* Groundsel. 2-9.
School-yard.

FLOWERING PLANTS.

3. *S. sylvaticus.* Wood Senecio. 7-9.
Egginton Road.

5. *S. aquaticus.* Water Senecio. 7-9.
Old Trent, near Bull's Meadow.

6. *S. Jacobœa.* Ragwort. 7-9.
Tan-yard Lane.

7. *S. erucifolius.* Narrow-leaved Senecio. 7-8.
Dry-banks.

XIX.—BIDENS.

1. *B. cernua.* Bur-Marigold, 7-9.
Crewe's Pond.

2. *B. tripartita.* Three-cleft Bidens. 7-9.
Brook-end.

XXI.—ARCTIUM.

1. *A. Lappa.* Bur-dock. 7-8.
Tan-yard Lane.

XXII.—SERRATULA.

1. *S. tinctoria.* Common Sawwort. 8.
Groves and Pastures.

XXIV.—CARDUUS.

2. *C. nutans.* Musk Thistle. 6-8.
Milton Road.

3. *C. acanthoides.* Welted Thistle. 6-7.
Willington Road.

5. *C. lanceolatus.* Spear Thistle. 7-9.
Burton Road.

6. *C. palustris.* Marsh Thistle. 7-8.
Osier-beds, near Twyford Road.

7. *C. arvensis.* Creeping Thistle. 7.
Fields near Foremark.

11. *C. pratensis.* Meadow Thistle. 7.
Low Meadow, adjoining Willington Junction.

XXV.—ONOPORDON.

1. *O. Acanthium.* Scotch Thistle. 7-8.
Burton Road.

XXVI.—CARLINA.

1. *C. vulgaris.* Common Carline. 7-9.
Ticknall Quarry.

XXVII.—CENTAUREA.

1. *C. nigra.* Knapweed. 6-8.
Bathing-field.

2. *C. Scabiosa.* Greater Centaurea. 7-8.
Ticknall Quarry.

3. *C. Cyanus.* Corn-flower. 7-8.
Corn-fields.

XXVIII.—TRAGOPOGON.

1. *T. pratense.* Yellow Goat's-beard. 6-7.
Near the Hall Orchard and Cricket-field Wall.

XXIX.—HELMINTHIA.

1. *H. echioides.* Oxtongue Helminth. 6-7.
Hedges and borders of fields.

XXX.—PICRIS.

1. *P. hieracioides.* Hawkweed Picris. 7-9.
Dry banks and borders of fields.

XXXI.—LEONTODON.

1. *L. hispidus.* Common Hawkbit. 6-9.
Milton Road.
2. *L. autumnalis.* Autumnal Hawkbit. 7-8.
Wall between Market-cross and Hall Orchard.
3. *L. hirtus.* Lesser Hawkbit. 7-9.
Milton Road.

XXXII.—HYPOCHŒRIS.

2. *H. radicata.* Cat's-ear. 7-8.
Milton Road.

XXXIII.—LACTUCA.

1. *L. muralis.* Wall Lettuce. 7-9.
Anchor Church.

XXXIV.—SONCHUS.

1. *S. arvenis.* Corn Sowthistle. 7-9.
Corn-fields on path to Bretby.
6. *S. oleraceus.* Common Sowthistle. 6-9.
Gravel-pit down Repton Village.

XXXV.—TARAXACUM.

1. *T. Dens-leonis.* Common Dandelion. 3-7.
Schoolyard.

XXXVI.—CREPIS.

3. *C. virens.* Smooth Crepis. 7-9.
Parson's Hills.

XXXVII.—HIERACIUM.

1. *H. Pilosella.* Mouse-ear Hawkweed. 6-7.
Old wall facing Cricket-field Brook.

3. *H. vulgatum.* Wall Hawkweed. 7-8.
   Cricket-field Wall.

5. *H. umbellatum.* Umbellate Hawkweed. 7-9.
   Between Crewe's Pond and Loscoe.

6. *H. sabaudum.* Savoy Hawkweed. 7-9.
   On the way to Repton Shrubs.

XXXVIII.—CICHORIUM.

1. *C. Intybus.* Succory. 6-7.
   Cornfield near Repton Lodge.

XL.—LAPSANA.

1. *L. communis.* Nipplewort. 7-8.
   Near the Vicarage.

ORDER XLIV.—CAMPANULACEÆ.

II.—JASIONE.

1. *J. montana.* Sheep's-bit. 7-8.
   Parson's Hills.

IV.—CAMPANULA.

3. *C. latifolia.* Giant Campanula. 7-8.
   Repton Shrubs.

7. *C. rotundifolia.* Harebell. 7-9.
   Cricket-field Wall.

9. *C. hybrida.* Corn Campanula. 6-9.
   Field between Milton and Repton Rocks.

ORDER XLV.—ERICACEÆ.

I.—VACCINIUM.

1. *V. Myrtillus.* Bilberry. 5.
Repton Rocks.

VII.—ERICA.

1. *E. vulgaris.* Common Heath. 8-9.
Repton Rocks.
2. *E. cinerea.* Scotch Heath. 7-8.
Repton Rocks.
3. *E. Tetralix.* Cross-leaved Heath. 7-8.
Repton Rocks.

ORDER XLVI.—PRIMULACEÆ.

I.—HOTTONIA.

1. *H. palustris.* Water Violet. 6.
Swamps on both sides of the road near Willington Bridge.

II.—PRIMULA.

1. *P. veris.* Common Primrose. 3-5.
Repton Shrubs (Primrose). Meadows (Cowslip). Repton Shrubs (Oxlip).

IV.—LYSIMACHIA.

3. *L. Nummularia.* Moneywort. 6-7.
Green Lane.
2. *L. nemorum.* Wood Lysimachia. 6-7.
Repton Shrubs and Repton Rocks.

4

VII.—ANAGALLIS.

1. *A. arvensis.* Common Pimpernel. 6-7.
Willington Road.

2. *A. tenella.* Bog Pimpernel. 7-8.
Boggy ground at Repton Rocks.

ORDER XLVII.—LENTIBULACEÆ.

II.—UTRICULARIA.

1. *U. vulgaris.* Common Bladderwort. 6-7.
Old Trent by the Bathing-field Lane.

ORDER XLVIII.—JASMINACEÆ.

I.—FRAXINUS.

1. *F. excelsior.* Common Ash. 4-5.
Field beyond the Stein-yard.

II.—LIGUSTRUM.

1. *L. vulgare.* Common Privet. 5-6.
Hedges on Willington Road.

ORDER L.—GENTIANACEÆ.

II.—ERYTHRÆA.

1. *E. Centaurium.* Centaury. 5-7.
Marshes by Willington Junction.

III.—GENTIANA.

5. *G. campestris.* Field Gentian. 7-8.
Field above Repton Shrubs.

V.—MENYANTHES.

1. *M. trifoliata.* Buckbean. 6-7.

ORDER LII.—CONVOLVULACEÆ.

I.—CONVOLVULUS.

1. *C. arvensis.* Lesser Convolvulus. 6-7.
Askew Hill.

2. *C. sepium.* Larger Convolvulus. 7-9.
Osier-bed next Willington Road.

II.—CUSCUTA.

3. *C. Epithymum.* Lesser Dodder. 6-8.
C. trifolii. Clover Dodder.
Corn-fields, Bretby-in-Clover.

ORDER LIII.—BORAGINEÆ.

IV.—LITHOSPERMUM.

1. *L. arvense.* Corn Gromwell. 5-7.
Road to Egginton.

2. *L. officinale.* Common Gromwell. 6-7.
Corn-fields.

V.—MYOSOTIS.

1. *M. palustris.* Forget-me-not. 6-10.
Old Trent.

3. *M. arvensis.* Field Myosote. 6-8.
Askew Hill Fields.

4. *M. collina.* Early Myosote. 4-5.
Askew Hill and Anchor Church.

5. *M. versicolor.* Changing Myosote. 4-6.
Askew Hill and Parson's Hills.

VII.—LYCOPSIS.

1. *L. arvensis.* Small Bugloss. 6-8.
   Between Milton and Repton Rocks.

VIII.—SYMPHYTUM.

1. *S. officinale.* Common Comfrey. 5-8.
   Osier-beds near the Stein-yard.

XI.—CYNOGLOSSUM.

1. *C. officinale.* Common Hound's-tongue. 6-8.
   By the lake at Calke Abbey.

ORDER LIV.—SOLANACEÆ.

I.—DATURA.

1. *D. Stramomium.* Thorn Apple. 9-10.
   Found in the Gravel of the Chapel Yard.

III.—SOLANUM.

1. *S. Dulcamara.* Bittersweet. 6-7.
   Parson's Hills.
2. *S. nigrum.* Black Nightshade. 7-10.
   Corner of Cricket-field.

IV.—ATROPA.

1. *A. Belladonna.* Deadly Nightshade. 6-8.
   In the Garden of the Hall, remaining from the Priory Herbarium.

ORDER LV.—OROBANCHACEÆ.

I.—OROBANCHE.

1. *O. major.* Greater Broomrape. 6-7.
   Repton Rocks, near the Pond.

II.—LATHRÆA.

1. *L. squamaria.* Toothwort. 4-5.
Calke Park.

ORDER LVI.—SCROPHULARINEÆ.

I.—VERBASCUM.

1. *V. Thapsus.* Great Mullein. 7-8.
Ticknall Quarry.

2. *V. Blattaria.* Moth Mullein. 7-8.
Hedge on further side of Stein-yard Brook.

III.—LINARIA.

1. *L. vulgaris.* Toad Flax. 7-9.
Midland Railway, on both sides of Willington.

6. *L. Cymbalaria.* Ivy Linaria. 5-10.
Cricket-field Wall by the Brook.

IV.—SCROPHULARIA.

1. *S. nodosa.* Knotted Figwort. 6-7.
Willington and Burton Roads.

2. *S. aquatica.* Water Figwort. 7-8.
Brook End.

V.—MIMULUS.

1. *M. luteus.* Yellow Mimulus. 5-9.
Knoll Hills.

VI.—LIMOSELLA.

1. *L. aquatica.* Common Limosel. 6-7.
Marshy Ground behind Foremark Hall.

VIII.—DIGITALIS.

1. *D. purpurea.* Purple Foxglove. 6-8.
Repton Rocks and Anchor Church.

IX—VERONICA.

4. *V. serpyllifolia.* Thyme-leaved Veronica. 5-7. Parson's Hills.

5. *V. officinalis.* Common Veronica. 5-8. Repton Shrubs.

6. *V. Anagallis.* Water Veronica. 6-8. Brook by Cricket-field.

7. *V. Beccabunga.* Brooklime. 6-8. Brook by Cricket-field.

9. *V. montana.* Mountain Veronica. 5-6. Repton Rocks.

10. *V. Chamædrys.* Germander Veronica. 5-7. Willington Road.

11. *V. hederæfolia.* Ivy Veronica. 6-9. Corn-fields on Askew Hill.

12. *V. agrestis.* Procumbent Veronica. 5-9. Askew Hill.

13. *V. Buxbaumii.* Buxbaum's Veronica. 8-10.

14. *V. arvensis.* Wall Veronica. 4-9. Cricket-field Wall.

X.—BARTSIA.

3. *B. Odontites.* Red Bartsia. 8-9. Field between Milton and Repton Rocks.

XI.—EUPHRASIA.

1. *E. officinalis.* Common Eyebright. 7-8. Ticknall Quarry.

XII.—RHINANTHUS.

1. *R. Crista-galli.* Common Rattle. 6. Hay-fields.

XIII.—PEDICULARIS.

1. *P. palustris.* Red Rattle. 6-9. Repton Rocks.
2. *P. sylvatica.* Lousewort. 6-8. Repton Rocks.

XIV.—MELAMPYRUM.

3. *M. pratense.* Common Melampyre. 6-8. Wood near Repton Rocks.

ORDER LVII.—LABIATÆ.

I.—SALVIA.

1. *S. pratensis.* Meadow Sage. 7. Swarkestone.

II.—LYCOPUS.

1. *L. europæus.* Gipsy-wort. 7-8. Ditches.

III.—MENTHA.

1. *M. sylvestris.* Horse Mint. 8-9. Field next the Tanyard Osier-bed.
2. *M. rotundifolia.* Round-leaved Mint. 7-9. Waste-ground.
5. *M. aquatica.* Water Mint. 8-9. Brook by Cricket-field Wall.

FLOWERING PLANTS.

IV.—THYMUS.

1. *T. Serpyllum.* Wild Thyme. 6-9.
Gravel-pit between Crewe's Pond and Bretby.

V.—ORIGANUM.

1. *O. vulgare.* Wild Marjoram. 7-8.
Probably at Ticknall.

VI.—CALAMINTHA.

1. *C. Acinos.* Basil Thyme. 7-8.
Gravel-pit between Crewe's Pond and Bretby.
2. *C. officinalis.* Common Calamint. 7-8.
Waysides and Hedges.
3. *C. Clinopodium.* Wild Basil. 7-8.
Gravel-pit between Crewe's Pond and Bretby.

VII.—NEPETA.

1. *N. Glechoma.* Ground Ivy. 4-6.
Hedge of Hall Orchard.
2. *N. Cataria.* Cat Mint. 7-9.
Ticknall Quarry.

VIII.—PRUNELLA.

1. *P. vulgaris.* Self-heal. 7-8.
Bathing-field Lane.

IX.—SCUTELLARIA.

1. *S. galericulata.* Common Skullcap. 7-9.
Osier-beds. On the Canal-side.

XI.—MARRUBIUM.

1. *M. vulgare.* White Horehound. 7-9.
Waste-ground.

XII.—STACHYS.

1. *S. Betonica.* Betony Stachys. 7-8.
Bathing-field.

3. *S. sylvatica.* Hedge Stachys. 7-8.
Hedges on Parson's Hills.

4. *S. palustris.* Marsh Stachys. 7-8.
Osier-beds.

XIII.—GALEOPSIS.

3. *G. Tetrahit.* Hemp Nettle. 7-9.
Corn-fields near Willington.

XIV.—BALLOTA.

1. *B. nigra.* Black Ballota. 7-9.
Top of Tanyard Lane.

XVI.—LAMIUM.

1. *L. amplexicaule.* Henbit. 5-9.
Corn-field near Willington.

2. *L. purpureum.* Red Lamium. 5-10.
Fields on Askew Hill.

3. *L. album.* Dead Nettle. 5-9.
Twyford Road.

4. *L. maculatum.* Spotted Lamium. 5-9.
Corner of Cricket-field and Courtyard at Knoll Hills.

5. *L. Galeobdolon.* Archangel. 5-7.
Anchor Church and Repton Shrubs.

XVII.—TEUCRIUM.

1. *T. Scorodonia.* Wood Sage. 6-8.
Anchor Church.

XVIII.—AJUGA.

1. *A. reptans.* Creeping Bugle. 5-6.
Repton Shrubs.

ORDER LVIII.—VERBENACEÆ.
I.—VERBENA.

1. *V. officinalis.* Common Vervein. 7-8.
By the Lake at Calke Abbey.

ORDER LX.—PLANTAGINEÆ.
I.—PLANTAGO.

1. *P. major.* Greater Plantain. 6-7.
Under Churchyard Wall.

2. *P. media.* Hoary Plantain. 6-7.
Parson's Hills. Fields near Ticknall.

3. *P. lanceolata.* Ribwort Plantain. 6-7.
Willington Road.

5. *P. Coronopus.* Buck's-horn Plantain. 6-7.
Goods Station near Willington Junction.

SUB-CLASS IV.—MONOCHLAMYDEÆ.

ORDER LXI.—PARONYCHIACEÆ.
IV.—SCLERANTHUS.

1. *S. annuus.* Annual Knawel. 7-11.
Gravelly Corn-fields.

2. *S. perennis.* Perennial Knawel. 7-11.
Dry sandy fields.

ORDER LXII.—CHENOPODIACEÆ.

IV.—CHENOPODIUM.

2. *C. polyspermum.* Many-seeded Goosefoot. 7-8.
Road to Newton.

3. *C. album.* White Goosefoot. 7-9.
Corn-fields by Burton Road.

5. *C. rubrum.* Red Goosefoot. 8-9.
A weed of cultivation near walls.

6. *C. urbicum.* Upright Goosefoot. 8-9.
Chiefly near houses.

7. *C. murale.* Nettle-leaved Goosefoot. 8-9.
Chiefly near houses.

8. *C. hybridum.* Maple-leaved Goosefoot. 8-9.
A weed of cultivation.

9. *C. Bonus-Henricus.* Good King Henry. 7-9.
By the School Arch.

VI.—ATRIPLEX.

4. *A. patala.* Common Orache. 7-8.
Burton Road.

ORDER LXIII.—POLYGONACEÆ.

I.—RUMEX.

2. *R. crispus.* Curled Dock. 7.
By Old Trent.

3. *R. obtusifolius.* Broad Dock. 7-8.
Roadsides.

4. *R. Hydrolapathum.* Water Dock. 7-8.
Old Trent by the Culvert.

5. *R. conglomeratus.* Clustered Dock. 7-8.

6. *R. sanguineus.* Red-veined Dock. 7-8.
By Twyford Lane.

9. *R. Acetosa.* Sorrel Dock. 6-7.
Tanyard Lane.

10. *R. Acetosella.* Sheep-sorrel Dock. 5-7.
Cricket-field Wall.

III.—POLYGONUM.

1. *P. aviculare.* Knotgrass. 7-8.
Burton Road.

3. *P. Convolvulus.* Climbing Buckwheat. 7-8.
Fields on Askew Hill.

6. *P. Bistorta.* Bistort. 6.
Osier-beds near Tanyard.

7. *P. amphibium.* Amphibious Polygonum. 7-9.
Crewe's Pond and Old Trent by Twyford Lane.

8. *P. Persicaria.* Common Persicaria. 7-8.
Brook-end, etc.

10. *P. Hydropiper.* Water-pepper Polygonum. 7-8.
Burton Road and Brook down the Village.

ORDER LXIV.—THYMELEACEÆ..

I.—DAPHNE.

2. *D. Laureola.* Spurge Laurel. 3.
Foremark.

ORDER LXVIII.—EUPHORBIACEÆ.

I.—EUPHORBIA.

2. *E. Helioscopia.* Sun Spurge. 7-8.
Burton Road.

6. *E. Peplus.* Petty Spurge. 6-9.
Corn-fields.

7. *E. exigua.* Dwarf Spurge. 6-9.
Askew Hill.

II.—MERCURIALIS.

1. *M. perennis.* Dog's Mercury. 4-5.
Anchor Church.

ORDER LXX.—CALLITRICHACEÆ.

II.—CALLITRICHE.

1. *C. aquatica.* Common Callitriche. 7-9.
Osier-beds below Crewe's Pond.

ORDER LXXI.—URTICACEÆ.

I.—URTICA.

1. *U. urens.* Small Nettle. 7-9.
By the Tanyard.

3. *U. dioica.* Common Nettle. 7-9.
Upper Paddock.

II.—PARIETARIA.

1. *P. officinalis.* Wall Pellitory. 7-9.
Cricket-field Wall.

III.—HUMULUS.

1. *H. Lupulus.* Common Hop. 7-9.
Hedge on the further side of the Stein-yard Brook.

ORDER LXXII.—ULMACEÆ.

I.—ULMUS.

1. *U. montana.* Wych Elm. 3. School-yard.
2. *U. campestris.* Common Elm. 3. School-yard.

ORDER LXXIII.—AMENTACEÆ.

II.—ALNUS.

1. *A. glutinosus.* Alder. 4. By Old Trent.

III.—BETULA.

1. *B. alba.* Birch. 5. Foremark Park.

IV.—CARPINUS.

1. *C. Betulus.* Hornbeam. 4.

V.—CORYLUS.

1. *C. Avellana.* Hazel. 3-4. Repton Shrubs.

VI.—FAGUS.

1. *F. sylvatica.* Beech. 4-5. Parson's Hills.

VII.—QUERCUS.

1. *Q. Robur.* British Oak. 4-5. Twyford Lane.

VIII.—SALIX.

1. *S. pentandra.* Bay Willow. 4-5.
2. *S. fragilis.* Crack Willow. 4.

3. *S. alba.* Common Willow. 4.
Near the Stein-yard.

4. *S. amygdalina.* Almond Willow. 4.
Between Milton and Foremark.

5. *S. purpurea.* Purple Willow. 3-4.
Near Willington.

6. *S. viminalis.* Common Osier. 4-5.
Tanyard Osier-beds.

7. *S. Caprea.* Common Sallow. 3-4.

8. *S. aurita.* Round-eared Willow. 3-4.

10. *S. repens.* Creeping Willow. 4-5.
Between Willington and Twyford.

IX.—POPULUS.

1. *P. alba.* White Poplar. 4.
Tanyard Lane.

2. *P. tremula.* Aspen Poplar. 3-4.

3. *P. nigra.* Black Poplar. 3-4.

ORDER LXXIV.—CONIFERÆ.

I.—PINUS.

1. *P. sylvestris.* Scotch Fir. 4-5.
Repton Park.

III.—TAXUS.

1. *T. baccata.* Yew. 4-5.
Foremark.

CLASS II.
MONOCOTYLEDONES.

SUB-CLASS I.—PETALOIDEÆ.

ORDER LXXV.—TYPHACEÆ.

I.—TYPHA.

1. *T. latifolia.* Great Bulrush. 7-9.
Swamps by the Railway Lines.

2. *T. angustifolia.* Lesser Bulrush. 7-9.
Ponds.

II.—SPARGANIUM.

1. *S. ramosum.* Branched Bur-reed. 7-8.
Old Trent, near Culvert.

ORDER LXXVI.—AROIDEÆ.

I.—ARUM.

1. *A. maculatum.* Cuckoo-pint. 5-6.
Parson's Hills.

II.—ACORUS.

1. *A. calamus.* Sweet Sedge. 6-7.
Near Willington Bridge and the Junction Osier-beds.

ORDER LXXVII.—LEMNACEÆ.

I.—LEMNA.

1. *L. trisulca.* Ivy-leaved Duckweed. 6-8.
Near Willington, by the roadside.

2. *L. minor.* Lesser Duckweed. 5-8. Swamps near Willington Bridge.

3. *L. gibba.* Gibbons Duckweed. 6-9. Common.

4. *L. polyrrhiza.* Greater Duckweed. Common.

ORDER LXXVIII.—NAIADEÆ.

III.—ZANNICHELLIA.

1. *Z. palustris.* Horned Pondweed.

V.—POTOMAGETON.

1. *P. natans.* Broad Pondweed. 6-8. Old Trent and the Millstream.

3. *P. lucens.* Shining Pondweed. 6-8. Old Trent.

5. *P. perfoliatus.* Perfoliate Pondweed. 6-8. Millstream.

6. *P. crispus.* Curly Pondweed. 6-7. Millstream.

7. *P. densus.* Opposite Pondweed. 7-8. Old Trent.

9. *P. pectinatus.* Fennel Pondweed. 7-8.

ORDER LXXIX.—ALISMACEÆ.

I.—BUTOMUS.

1. *B. umbellatus.* Flowering Rush. 6-7. New Trent, by Bathing-place.

II.—SAGITTARIA.

1. *S. sagittifolia.* Common Arrowhead. 7-9.
Old Trent, opposite Bull's Farm.

III.—ALISMA.

1. *A. Plantago.* Water Plantain. 7-8.
Old Trent, by Culvert.

2. *A. ranunculoides.* Lesser Alisma. 7-9.

VI.—TRIGLOCHIN.

1. *T. palustre.* Arrow-grass. 6-9.
Canal near Willington.

ORDER LXXXI.—ORCHIDACEÆ.

IV.—EPIPACTIS.

1. *E. latifolia.* Broad Epipactis. 6.
Repton Shrubs.

VI.—LISTERA.

1. *L. ovota.* Twayblade. 6-8.
Tanyard Osier-beds.

VII.—NEOTTIA.

1. *N. Nidus-avis.* Bird's Nest Neottia. 4-6.
Repton Shrubs.

IX.—SPIRANTHES.

1. *S. autumnalis.* Lady's Tresses. 9-10.

X.—ORCHIS.

1. *O. Morio.* Green-winged Orchis. 5-6.
Parson's Hills.

5. *O. mascula.* Early Orchis. 4-6.
End of Parson's Hills.

7. *O. macualata.* Spotted Orchis. 4-6.
Parson's Hills and swamps between canal and railway.

8. *O. latifolia.* Marsh Orchis. 4-6.
Parson's Hills.

XII.—HABENARIA.

1. *H. bifolia.* Butterfly Habenaria. 5-8.
Farm at Milton.

2. *H. albida.* Small Habenaria. 6-7.
Tanyard Osier-bed.

3. *H. viridis.* Green Habenaria. 6-7.

ORDER LXXXII.—IRIDEÆ.

I.—IRIS.

1. *I. Pseudacorus.* Yellow Flag. 6-7.
Stein-yard.

V.—CROCUS.

2. *C. nudiflorus.* Naked Crocus. 8-9.
Meadows by the Trent.

ORDER LXXXIII.—AMARYLLIDEÆ.

I.—NARCISSUS.

1. *N. Pseudo-narcissus.* Daffodil. 3.
Osier-bed by Old Trent.

II.—GALANTHUS.

1. *G. nivalis.* Snowdrop. 2-3.
Osier-bed by Old Trent.

ORDER LXXXIV.—DIOSCORIDEÆ.

I.—TAMUS.

1. *T. communis.* Black Bryony. 5-7.
Hedges by the Bathing-field.

ORDER LXXXV.—LILIACEÆ.

I.—PARIS.

1. *P. quadrifolia.* Herb Paris. 5.
Repton Rocks.

III.—CONVALLARIA.

1. *C. majalis.* Lily of the Valley. 5.
Anchor Church.

VI.—RUSCUS.

1. *R. aculeatus.* Butcher's Broom. 3-4.
Smith's Plantation.

XI.—ORNITHOGALUM.

1. *O. umbellatum.* Star of Bethlehem. 4-5.
Found by Dr. Pears near the Hall.

XII.—SCILLA.

3. *S. nutans.* Bluebell. 5-6.
Parson's Hills.

XIV.—ALLIUM.

7. *A. ursinum.* Ramsons. 5-6.
Repton Shrubs.

XVIII.—COLCHICUM.

1. *C. autumnale.* Meadow Saffron. 9-10.
Meadows by the Trent at Anchor Church.

ORDER LXXXVI.—JUNCACEÆ.

I.—JUNCUS.

1. *J. communis.* Common Rush. 7-8.
Old Trent.

2. *J. glaucus.* Hard Rush. 7-8.
By the Old Trent.

5. *J. articulatus.* Jointed Rush. 7-8.
At the Junction of Old and New Trent.

9. *J. bufonius.* Toad Rush. 6-8.
Field next the Culvert.

II.—LUZULA.

1. *L. pilosa.* Hairy Woodrush. 5-6.
Repton Shrubs.

2. *L. sylvatica.* Great Woodrush. 5-6.
Anchor Rocks.

4. *L. campestris.* Field Woodrush. 3-5.
Stein-yard.

SUB-CLASS II.—GLUMACEÆ.

ORDER LXXXVIII.—CYPERACEÆ.

VI.—SCIRPUS.

1. *S. palustris.* Creeping Scirpus. 6-8.
Canal.

3. *S. multicaulis.* Many-stalked Scirpus. 6-7.
Wet ground by the Old Trent.

7. *S. setaceus.* Bristle Scirpus. 6-7.
Junction of Old and New Trent.

12. *S. lacustris.* Lake Scirpus. 6-7.
Old Trent.

14. *S. sylvaticus.* Wood Scirpus. 6-7.
Osier-beds.

VII.—ERIOPHORUM.

3. *E. polystachum.* Common Cottonsedge.
Repton Rocks, in the Marsh.

IX.—CAREX.

5. *C. leporina.* Oval Carex. 5-6.
Repton Rocks.

8. *C. stellulata.* Star-headed Carex. 5-6.
Repton Rocks, near the Wood.

9. *C. canescens.* Whitish Carex. 5-6.
Near Bull's Farm.

10. *C. remota.* Remote Carex. 5-6.
Crewe's Pond.

11. *C. axillaris.* Axillary Carex. 6.
Banks of Canal.

12. *C. paniculata.* Panicled Carex. 6.
Boggy Ground.

13. *C. vulpina.* Fox Carex. 6.
River Banks.

14. *C. muricata.* Prickly Carex. 5-6.
Near the New Trent.

15. *C. arenaria.* Sand Carex. 6-7.
Repton Rocks.

19. *C. cæspitosa.* Tufted Carex. 4-6.
Osier-beds near Bull's Farm.

20. *C. acuta.* Acute Carex. 4-6.
By the Trent.

26. *C. præcox.* Vernal Carex. 4-5.
Stein-yard.

28. *C. pilulifera.* Pill-headed Carex. 5.
Repton Rocks.

31. *C. hirta.* Hairy Carex. 4-6.
Osier-bed by Junction of Old and New Trent.

34. *C. flava.* Yellow Carex. 4-7.
Repton Rocks.

35. *C. distans.* Distant Carex. 6-7.
Repton Rocks.

37. *C. panicea.* Carnation Grass. 5-6.
Marshes.

40. *C. glauca.* Glaucous Carex. 5-6.
Frequent.

41. *C. sylvatica.* Wood Carex. 5-6.
Repton Shrubs.

43. *C. Pseudocyperus.* Cyperus-like Carex. 5-6.
Marsh at Willington Junction.

44. *C. pendula.* Pendulous Carex. 5-6.
Repton Shrubs.

45. *C. ampullacea.* Bottle Carex. 5-6.
Near Bull's Farm.

47. *C. paludosa.* Marsh Carex. 4-6.
Banks of Old Trent.

ORDER LXXXIX.—GRAMINEÆ.

II.—MILIUM.

1. *M. effusum.* Spreading Milium. 6-7.
Woods.

V.—ANTHOXANTHUM.

1. *A. odoratum.* Vernal Grass. 5-6.
Pastures.

VI.—PHALARIS.

1. *P. canariensis.* Canary Phalaris. 6-7.
Bathing-field.

VII.—DIGRAPHIS.

1. *D. arundinacea.* Reed Digraphis. 6-7.
Field next Bathing-field.

VIII.—PHLEUM.

1. *P. pratense.* Timothy-grass. 5-6.
Bank of Trent, opposite Bathing-field.

IX.—ALOPECURUS.

1. *A. agrestis.* Slender Foxtail. 5-8.
Fields and Waysides.

2. *A. pratensis.* Meadow Foxtail. 5-8.
Everywhere.

3. *A. geniculatus.* Marsh Foxtail. 5-8.
Old Trent at Culvert.

XIII.—AGROSTIS.

1. *A. alba.* Common agrostis. 5-8. Everywhere.
2. *A. canina.* Brown Agrostis. 6-7. Scaddow Rocks.

XVII.—AIRA.

1. *A. cæspitosa.* Tufted Aira. 6-7. Marshy ground.
2. *A. flexuosa.* Wavy Aira. 6-7. Wood at Repton Rocks.
4. *A. præcox.* Early Aira. 4-5. Sandy hills and pastures.
5. *A. caryophyllea.* Hair-grass. 5-6. Gravel pit down the Village.

XVIII.—AVENA.

1. *A. fatua.* Wild Oat. 6-7. Corn-fields.
2. *A. pratensis.* Perennial Oat. 5-6. Grass-fields.
3. *A. flavescens.* Yellow Oat. 6-7. Dry Meadows.

XIX.—ARRHENATHERUM.

1. *A. avenaceum.* Common False Oat. 6-7. Bathing-field Lane.

XX.—HOLCUS.

1. *H. lanatus.* Common Holcus. 5-8. Fields.
2. *H. mollis.* Soft Holcus. 6-7. Fields.

XXVI.—HORDEUM.

2. *H. pratense.* Meadow Barley. 5-6. Dryfields.

3. *H. murinum.* Wall Barley. 6-7. Churchyard Wall.

XXVII.—TRITICUM.

1. *T. repens.* Couchgrass. 6-7. Fields and Waste Places.

2. *T. caninum.* Fibrous Triticum. 6-7. Woods and Banks.

XXVIII.—LOLIUM.

1. *L. perenne.* Ryegrass Lolium. 5-8. Bathing-field.

2. *L. temulentum.* Darnel Lolium. 6-7. Parson's Hills.

XXX.—BROMUS.

2. *B. asper.* Hairy Brome. 6-7. Woods.

3. *B. sterilis.* Barren Brome. 5-6. Upper Paddock.

6. *B. arvensis.* Field Brome. 4-7. Everywhere.

7. *B. giganteus.* Tall Brome. 6-7.

XXXI.—FESTUCA.

1. *F. ovina.* Sheep's Fescue. 6-7. Pastures.

2. *F. elatior.* Meadow Fescue. 5-6. Meadows and moist pastures.

4. *F. Myurus.* Rat's-tail Fescue. 5-6.
Repton Park Wall.

XXXII.—DACTYLIS.

1. *D. glomerata.* Clustered Cock's-foot. 5-8.
Fields and Roadsides.

XXXIII.—CYNOSURUS.

1. *C. cristatus.* Crested Dog's-tail. 6-7.
Fields and Roadsides.

XXXIV.—BRIZA.

1. *B. media.* Quakegrass. 5-6.
Fields.

XXXV.—POA.

1. *P. aquatica.* Reed Poa. 6-7.
Marshy ground.

2. *P. fluitans.* Floating Poa. 5-8.
Old Trent at the Culvert.

5. *P. procumbens.* Procumbent Poa. 6-7.

6. *P. rigida.* Hard Poa. 6-7.
Ticknall Quarry.

8. *P. annua.* Annual Poa. 4-9.
Stein-yard.

10. *P. pratensis.* Meadow Poa. 5-7.
Meadows and pastures.

11. *P. trivialis.* Roughish Poa. 5-7.
Meadows and pastures.

12. *P. nemoralis.* Wood Poa. 6-7.
Repton Rocks.

XXXVIII.—MELICA.

2. *M. uniflora.* Wood Melick. 5-6. Loscoe Hill.

XLII.—ARUNDO.

1. *A. Phragmites.* Common Reed. 7-9. Watersides.

CLASS III.—ACOTYLEDONES.

ORDER XCI.—EQUISETACEÆ.

I.—EQUISETUM.

1. *E. Telmateia.* Great Equisetum. 4-5. Repton Rocks.

2. *E. arvense.* Common Horsetail. 4-5. Bathing-field Lane.

3. *E. sylvaticum.* Wood Equisetum. 5-6. Calke Park and Repton Shrubs.

5. *E. limosum.* Smooth Equisetum. 6-7. Crewe's Pond.

6. *E. palustre.* Marsh Equisetum. 6-7. Stagnant Water.

7. *E. hyemale.* Rough Equisetum. 6-8. Repton Rocks.

ORDER XCII.—FILICES.

I.—OPHIOGLOSSUM.

1. *O. vulgatum.* Common Adder's-tongue. 6-7. Near Foremark and Bathing-field.

FLOWERING PLANTS.

II.—BOTRYCHIUM.

1. *B. Lunaria.* Common Moonwort. 4-5.
   Sandy fields near Foremark Hall and Southwood Close.

IV.—POLYPODIUM.

1. *P. vulgare.* Common Polypody. 5-10.
   School Walls.

VII.—ASPIDIUM.

2. *A. aculeatum.* Prickly Shieldfern. 6-8.
   Bretby-lane.
3. *A. Thelypteris.* Marsh Shieldfern. 6-8.
   Repton Rocks.
4. *A. Oreopteris.* Mountain Shieldfern. 6-8.
   Repton Shrubs.
5. *A. Filix-mas.* Male Shieldfern. 5-9.
   Tanyard-lane.
7. *A. spinulosum.* Broad Shieldfern. 5-9.
   Repton Rocks.

VIII.—ASPLENIUM.

1. *A. Filix-fœmina.* Lady Spleenwort. 5-9.
   Repton Shrubs.
5. *A. Trichomanes.* Common Spleenwort. 4-11.
   Anchor Church.
6. *A. Adiantum-nigrum.* Black Spleenwort. 5-10.
   Cricket-field Wall, Anchor Church.
8. *A. Ruta-muraria.* Wallrue Spleenwort. 5-10.
   Cricket-field Wall.

IX.—SCOLOPENDRIUM.

1. *S. vulgare.* Common Hart's-tongue. 5-9.
Cricket-field Wall.

XI.—BLECHNUM.

1. *B. Spicant.* Hard Blechnum. 6-9.
Repton Rocks.

XII.—PTERIS.

1. *P. aquilina.* Brake Pteris. 7-9.
Repton Shrubs.

ADDENDA.

The following plants have been found in the Repton district since Mr. Lewis's list was written; or, having been omitted by him, are re-introduced, on good authority, from the previous edition.

In giving the names, the seventh edition of the London Catalogue has been followed.

RANUNCULUS.
R. circinatus. Sibth.
River Trent, and Brook near Bull's in the Meadow. (W. G. and J. H.)

R. fluitans. Linn.
River Trent and Willington Canal. (W. G. and J. H.)

R. Drouetii. Schultz.
Twyford Brook. (W. G.)

ERANTHIS.
E. hyemalis. Salisb.
Plantations—not truly indigenous. (W. G.)

AQUILEGIA.
A. vulgaris. Linn.
Near Willington—probably an escape. (J. H.)

DELPHINIUM.
D. Ajacis. Reich.
Willington—probably an escape. (J. H.)

VIOLA.
V. odorata var. alba.
Bretby, Newton, etc. (W. G. and J. H.)

CERASTIUM.

C. triviale. Link.
Twyford Meadows—common.

C. semidecandrum. Linn.
Old walls.

SAGINA.

S. apetala. Linn.
Walls—common.

LOTUS.

L. major. Scop.
Milton and other places.

PRUNUS.

P. insititia. Linn.
Bretby.

ARTEMISIA.

A. Absinthium. Linn.
Egginton. (W. G.)

SCUTELLARIA.

S. minor. Linn.
Repton Rocks. (J. H.)

RUMEX.

R. pulcher. Linn.
Burton. (W. G.)

R. pratensis. M. and K.
Road between Milton and Repton Rocks. (W. G.)

POTAMOGETON.

P. prælongus. Wulf.
Willington Canal. (W. G. and J. H.)

P. zosterifolius. Schum.
Willington Canal. (W. G.)

P. pusillus. Linn.
Brook near Bull's in the Meadow. (J. H.)

ELODIA.

E. canadensis. Mich.
Ponds and streams—common. (W. G. and J. H.)

ALLIUM.

A. oleraceum. Linn.
Near Ingleby and Knoll Hills. (J. H.)

NEPHRODIUM.

N. dilatatum. Desv.
Common. (W. G.)

A LIST OF MOSSES.

ORDER II.—SPHAGNACEÆ.

II.—SPHAGNUM. Bog Moss.

S. cymbifolium. Dill. Ehrh. 6-7.
Bogs. Repton Rocks.

S. acutifolium. Ehrh. 6-7.
Bogs and marshes. Repton Rocks.

S. cuspidatum. Dill. Ehrh. 6-7.
Bogs. Repton Rocks.

S. squarrosum. Persoon. 6-7.
Bogs. Repton Rocks.

ORDER III.—BRYACEÆ.

SECTION I.—ACROCAPI.

SUB-ORDER I.—PHASCEÆ.

IV.—PHASCUM.—Earth Moss.

P. cuspidatum. Schreb. 3.
Banks and fallow-fields. Canada Gardens.

P. subulatum. Lin. Spring.
Fallow-fields, etc. Spurs Bottoms and Robin's Cross.

P. alternifolium. Br. and Schimp. Spring.
Banks, fallow-fields. Banks near Ingleby.

SUB-ORDER II.—WEISSIEÆ.

VI.—WEISSIA.

W. controversa. Hedw. Spring.
Banks, common. Between Park Pond and Bretby.

W. cirrhata. Hedw. Spring.
On walls in and about Repton.

SUB-ORDER V.—DICRANEÆ. Fork Mosses.

XVI.—DICRANUM.

D. cerviculatum. Hedw. 6-7.
Banks and rocks. Repton Rocks.

D. heteromallum. Hedw. 11-12.
Moist banks, very common. Roots of trees at Park Pond.

XVII. CERATODON.

C. purpureus. Brid. 4-5.
Walls and banks, common.

SUB-ORDER VI.—CAMPYLOPODEÆ.

XX—CAMPYLOPUS. Swan-neck Moss.

C. flexuosus. Dill., Lin. 11.
Moist, shady rocks. Repton Rocks.

SUB-ORDER VII.—POTTIEÆ.

XXI.—POTTIA.

P. truncata. Br. and Schimp. 2-3.
In fallow-fields and on banks.

SUB-ORDER VIII.—TRICHOSTOMEÆ.

XXV.—DIDYMODON.

D. rubellus. Br. and Schimp. 10.
On walls, rocks, etc. Shobnall.

XXVI.—TRICHOSTOMUM.

T. mutabile. Bruch. 6-7.
Shady banks, moist rocks, etc. Near Park Pond.

XXVII.—TORTULA. Screw Moss.

T. ambigua. Br. and Schimp. 11-12.
On walls, marly banks, etc. Shobnall.

T. unguiculata. Hedw. 12.
Banks and hedges, frequent. About Park Pond, etc

T. Hornschuchiana. Schultz. 4-5.
Walls, rocks, and marly banks. Shobnall.

T. muralis. Timm. 4-5.
Walls and stones. Very common in and about Repton.

T. ruralis. Hedw. 3-4.
On walls, roofs of buildings, etc., frequent.

SUB-ORDER X.—ENCALYPTEÆ.

XXIX.—ENCALYPTA. Extinguisher Moss.

E. vulgaris. Hedw. 3-4.
Walls, banks, and rocks, Park Pond Wall.

SUB-ORDER XII.—GRIMMIEÆ.

XXXIII.—GRIMMIA.

G. pulvinata. Smith. 3-4.
On walls, rocks, etc. Common on walls all about Repton.

SUB-ORDER XIV.—ORTHOTRICHEÆ.

XXXVII.—ORTHOTRICHUM. Bristle Moss.

O. anomalum. Hedw. Spring.
Trees and rocks.

O. diaphanum. Schrad. 4.
Trees, walls, etc.

SUB-ORDER XVI.—TETRAPHIDEÆ.

XXXIX.—TETRAPHIS. Four-tooth Moss.

T. pellucida. Hedw. 8-9.
Shady, rocky places, roots of dying trees. Repton Rocks.

SUB-ORDER XVIII.—POLYTRICHEÆ. Hair Mosses.

XLIII.—ATRICHUM.

A. undulatum. P. Beauv. 10-11.
Banks and woods—common. About Park Pond and Bretby Road.

XLV.—POGONATUM.

P. nanum. Bridel. 10-11.
Sandy shaded banks. Near Park Pond.

P. aloides. Bridel. 10-11.
Moist banks. Repton Rocks.

P. urnigerum. Bridel. 10-11.
Banks and sides of streams. Near Repton Rocks.

XLVI.—POLYTRICHUM.

P. formosum. Hedw. 6.
In woods, etc. "Dark Walk," Foremark Hall.

P. commune. Linn. 6.
In marshy and turfy places. Bog at Repton Rocks.

P. juniperinum. 5-6.
Walls, heaths, etc. At Ticknall and Foremark.

P. piliferum. Schreb. 5-6.
Walls and sandy ground. On wall going to Repton Rocks.

SUB-ORDER XIX.—BRYEÆ. THREAD MOSSES.

XLVIII.—AULACOMNION.

A. palustre. Schwaegr. 5-6.
Turfy bogs and marshes. Repton Rocks.

L.—LEPTOBRYUM.

L. pyriforme. Wilson. 5-6.
Sandstone Rocks. Cross Road from Newton to Bretby.

LI.—BRYUM.

B. nutans. Schreb. 5-6.
Sandy banks, etc. Repton Rocks.

B. annotinum. Hedw. 5-6.
Sandy banks, quarries, etc. Ticknall Lime Quarry.

B. *pallens.* Swartz. 6.
Moist places, ditches, etc. Willington Pits.

B. cernuum. Hedw. 5.
Walls, rocks, banks, etc. Foremark and Repton Rocks.

B. intermedium. Bridel. 6-12.
Walls and rocks. Long wall at Newton Solney.

B. cæspiticium. Linn. 5-6.
Walls and rocks. Common on walls all about Repton.

B. atropurpureum. Web. and Mohr. 5.
Banks, walls, etc. North Stafford Goods Station, Willington.

B. argenteum. Linn. 10-11.
Roadsides and walls. Willington, and near " Spread Eagle."

LII.—MNIUM. Thyme Thread Moss.

M. cuspidatum. Hedw. 3-4.
Shady rocks and walls. Anchor Church.

M. rostratum. Schwaegr. 4.
Moist, shady rocks and banks. Spurs Bottoms.

M. hornum. Linn. Hedwg. 5.
Shady banks and woods. Foremark.

M. undulatum. Hedw. 4-5.
Moist, shady banks. Hedge-banks at Bretby. Not in fruit.

SUB-ORDER XXII.—FUNARIEÆ.

LVIII.—FUNARIA. Cord Moss.

F. hygrometrica. Hedw. 4-11.
Banks, walls, etc., especially on burnt ground. About Bretby.

LX.—PHYSCOMITRIUM. Bladder Moss.

P. pyriforme. Br. and Schimp. 4.
Moist banks and ditches about Spurs Bottoms.

LXII.—BARTRAMIA. Apple Moss.

B. pomiformis. Hedw. 4.
Dry, shady banks. Robin's Cross.

SUB-ORDER XXVII.—SCHISTOSTEGEÆ.

LXXI.—SCHISTOSTEGA. Cavern Moss.

S. osmundacea. Web. and Mohr. 3.
In caves and crevices of sandstone rock. Repton Rocks.

SUB-ORDER XXVIII.—FISSIDENTEÆ.

LXXII.—FISSIDENS. Flat Fork Moss.

F. bryoides Hedw. 1-2.
Shady banks, etc., common. Burton Road.

F. taxifolius. Hedw. 12.
Moist, shady banks. Milton Road.

SECTION II.—PLEUROCARPI.

SUB-ORDER XXXI.—ISOTHECIÆ.

LXXIX.—ISOTHECIUM. Frond Moss.

I. myosuroides. Dill., Linn. 11.
Trunks of trees, rocks, and walls. Burton Road.

LXXX.—CLIMACIUM. Marsh Tree Moss.

C. dendroides. Web. and Mohr. 10.
Boggy or marshy places. Repton Rocks. Rare in fruit.

LXXXI.—CYLINDROTHECIUM. Cylinder Moss.

C. Montagnei. Bryol. Eur.
Rocks and hills, chiefly in limestone or chalk. Lime Quarry, Ticknall. Does not fruit in Britain.

SUB-ORDER XXXII.—HYPNEÆ.

LXXXIII.—HYPNUM. Feather Moss.

H. lutescens. Dill., Huds. 4.
Limestone rocks, sandy banks. The Oaks, Shobnall.

H. populeum. Swartz. 11-2.
Walls, rocks, and trees. Knoll Hills.

H. rutabulum. Dill., Linn. 11.
Banks, walls, and trees. Very common.

H. prælongum. Dill., Linn. 11.
Moist, shady banks. Near Bretby.

H. Swartzii. Turner. 11.
Moist banks and rocks. Hedge-banks near Bretby.

H. ruscifolium. Dill. 11.
Rocks and stones in brooks. Brook at Spurs Bottoms.

H. cuspidatum. Dill., Linn. 5-6.
In marshes. Willington Pits.

H. Schreberi. Dill. 10-11.
Woods and shady banks. Ticknall Lime Quarry. Rare in fruit.

H. purum. Dill., Linn. 10-11.
Shady banks. Robin's Cross.

H. tamariscinum. Hedw. 11.
Woods and banks. Banks near Bretby.

H. splendens. Dill., Sibth. 4.
Grassy banks, in woods, etc. Lime Quarry, Ticknall.

H. triquetrum. Dill., Linn. 11.
Woods, etc. Gravel pit, Repton.

H. squarrosum. Dill., Linn. 11.
Banks and woods. Banks near Bretby.

H. cupressiforme. Dill., Linn. 11-12.
Walls, rocks, trunks of trees, etc. Common.

H. undulatum. Dill., Linn. 4-5.
In woods, etc. Repton Rocks.

H. denticulatum. Dill., Linn. Summer.
In woods, hedge banks, etc. Common.

LXXXV.—NECKERA.

N. complanata. Bryol. Eur. 11-12.
On trunks of trees and walls. Shobnall.

SUB-ORDER XXXIV.—HOOKERIEÆ.

LXXXVI.—HOOKERIA.

H. lucens. Dill., Linn. 11-12.
Moist banks in woods, and among rocks. Repton Rocks.

SUB-ORDER XXXVI.—FONTINALEÆ.

LXXXIX.—FONTINALIS. Water Moss.

F. antipyretica. Linn. 6-7.
In water. River Trent, etc.

BIRDS

OF

REPTON AND THE NEIGHBOURHOOD.

(Compiled originally in 1865 and revised in 1881.)

ORDER—RAPTORES.
FAMILY—FALCONIDÆ.

Falco subbuteo. (Linn.) The Hobby.
  Very rare. Specimen in beautiful plumage shot by Mr. F. Holbrooke, Jun., within last few years (1881) in Bladon Wood.

Falco æsalon. (Genel.) The Merlin.
  Is occasionally shot in the neighbourhood.

Falco tinnunculus. (Linn.) The Kestrel.
  The commonest hawk of the neighbourhood; I have found two or three nests in a single day.

Accipiter nisus. (Linn.) The Sparrow Hawk.
  In spite of the wars waged against this bird, it is by no means of rare occurrence; it remains throughout the year.

Milvus vulgaris. (Linn.) The Kite or Glead.
  Is now very rarely seen—though it occurs sometimes at Dovedale, where it formerly has been known to build.

Buteo vulgaris. (Bechst.) The Buzzard.

Is now almost extinct here. Dr. Hewgill, however, can remember this species nesting in Repton Shrubs.

Circus cyaneus. (Fleming.) Hen Harrier.

Two eggs are in my possession, obtained by Mr. F. Drewry, at Drakelowe, about 1870.

FAMILY—STRIGIDÆ.

Strix otus. (Linn.) The Long-eared Owl.

Is now rare. The bird and eggs were taken in a fir plantation near Egginton Station some years ago.

Strix brachyotus. (Mont.) The Short-eared Owl.

Does not build, but is occasionally shot.

Strix flammea. (Linn.) The White Owl—local name, Barn Owl.

Builds with us. The tower of the church was formerly a favourite building-place.

Strix aluco. (Latham.) The Tawny Owl.

By no means common. Is occasionally met with in Bretby and Ingleby Woods.

Stryx nyctea. (Mont.) Snowy Owl.

Sir John Crewe records one killed near Burton-on-Trent.

ORDER—INSESSORES.

TRIBE—DENTIROSTRES.

FAMILY—LANIDÆ.

Lanius collurio. (Linn.) The Red-backed Shrike.

The eggs are sometimes taken; I have never had the

fortune to find a nest myself, though I have heard of several instances of this cruel little bird nidificating in Repton parish.

FAMILY—MUSCICAPIDÆ.

Muscicapa grisola. (Linn.) The Spotted Flycatcher.
Very common.

FAMILY—MERULIDÆ.

Cinclus aquaticus. (Bechst.) The Dipper.
Occasionally seen in the valley of the Dove. Has been seen at Newton Solney, at the junction of the Dove and Trent.

Turdus viscivorus. (Linn.) The Missel Thrush—local name, the Thrice Cock.

Turdus pilaris. (Linn.) The Fieldfare.
Common during the winter.

Turdus musicus. (Linn.) The Thrush—local name, the Throstle.

Turdus iliacus. (Linn.) The Redwing.
Of frequent occurrence throughout the winter months.

Turdus merula. (Linn.) The Blackbird.

Turdus torquatus. (Linn.) The Ring Ouzel.
One of a pair was shot on April 28th, 1848, at Newton Solney; also, Sir John Crewe records a specimen at Calke.

Oriolus galbula. (Linn.) The Golden Oriole.
Mr. Brown records the occurrence of this rare bird at Egginton in 1841.

FAMILY—SYLVIADÆ.

Sylvia modularis. (Lath.) The Hedge Sparrow.

Sylvia rubecula. (Pennant.) The Robin.

Sylvia phœnicurus. (Lath.) The Redstart—local name, the Firetail.

Sylvia rubicola. (Penn.) The Stonechat.
Rarely seen.

Saxicola rubetra. (Fleming.) The Winchat—local name, the Utic.
Nests with us in abundance, especially in the osier-beds near the Trent.

Saxicola œnanthe. (Fleming.) The Wheatear.
Very rare. (Found a nest in the bank of a ditch, in May, 1865. T. W. W.) Also occurs at Calke Park.

Salicaria locustella. (Selby.) The Grasshopper Warbler.
The above bird is of frequent occurrence in the breeding season; its eggs, however, are but seldom taken, owing to the skilful way in which it conceals its nest. Newton Brooks is a favourite resort of this species, I have in my possession two eggs taken from there in the spring of 1860.

Salicaria phragmitis. (Selby.) The Sedge Warbler.
Very common in the breeding season.

Salicaria arundinacea. (Selby.) The Reed Wren or Warbler.
By no means rare; it appears more common here than in most parts of the county. It especially resorts to the osier-beds.

Philomela luscinia. (Swains.) The Nightingale.
: This delightful songster but seldom pays us a visit. The last which I can remember sang for several weeks in succession in Bladon Wood, about 1861 or 1862.

Sylvia atricapilla. (Penn.) The Blackcap Warbler.
: I have frequently taken the eggs of this sweet songster; they vary much in shade, some being of a dark brown, while others are of a light pink.

Sylvia hortensis. (Lath.) The Garden Warbler.
: Is not so frequently met with as the last-mentioned species; the eggs are liable to the same variation of colour.

Sylvia cinerea. (Penn.) The Whitethroat—local name, Peggy. Very common.

Sylvia sylviella. (Penn.) The Lesser Whitethroat. Common.

Sylvia sylvicola. (Lath.) The Wood Wren.
: Not common. The nest is sometimes found at Foremark.

Sylvia trochilus. (Lath.) The Willow Wren—local name, Bank Jug. Very common.

Sylvia rufa. (Lath.) The Chiff Chaff.

Sylvia regulus. (Penn.) The Golden-crested Wren. Tolerably common.

FAMILY—PARIDÆ.

Parus major. (Linn.) The Great Tit—local name, Oxeye.

Parus cœruleus. (Linn.) The Blue Tit—local name, the Tom Tit.

Parus ater. (Linn.) The Cole Tit.

Parus palustris. (Linn.) The Marsh Tit.

Parus caudatus. (Linn.) The Long-tailed Tit—local name, Bottle Jug.

All the above remain with us during the year. The Marsh Tit is not often met with.

FAMILY—AMPELIDÆ.

Bombycilla garrula. (Flem.) The Bohemian Chatterer.

There are a few instances on record of the occurrence of this bird.

FAMILY—MOTACILLIDÆ.

Motacilla Yarrellii. (Gould.) The Pied Wagtail—local name, Water Wagtail.

Motacilla Boarula. (Penn.) The Grey Wagtail.

Not uncommon about Burton. Mr. J. T. Harris records a nest at Burton Mill within the last few years (1881).

Motacilla Rayi. (Schleg.) Ray's Wagtail—local name, Yellow Wagtail.

Breeds with us occasionally. The bird is common.

FAMILY—ANTHIDÆ.

Anthus arboreus. (Bechst.) The Tree Pipit—local name, the Tit Lark.

Anthus pratensis. (Bechst.) The Meadow Pipit.

Both are occasionally met with.

TRIBE—CONIROSTRES.
FAMILY—ALAUDIDÆ.

Alauda arvensis. (Linn.) The Sky Lark. Common.

Alauda arborea. (Linn.) The Wood Lark.
   This bird is rare, but it is met with sometimes at Calke Park.

FAMILY—EMBERIZIDÆ.

Emberiza miliaria. (Linn.) The Common Bunting. Not common.

Emberiza schœniclus. (Linn.) The Black-headed Bunting. Very common. Local name—the Blackcap.

Emberiza citrinella. (Linn.) The Yellow Hammer—local name, Goldie and Goldfinch

Emberiza cirlus. (Pennant.) The Cirl Bunting.
   Taken within last few years (1881) in Bladon Wood.

FAMILY—FRINGILLIDÆ.

Fringilla cœlebs. (Linn.) The Chaffinch—local name, Piedfinch.

Fringilla montifrigilla. (Pennant.) The Brambling.
   Sir John Crewe records this as not being uncommon.

Passer montanus. (Briss.) The Tree Sparrow.
   I have taken the eggs of this species in the willow trees near the Old Trent. It is, however, by no means common.

Passer domesticus. (Briss.) The House Sparrow—local name, Eaves Sparrow.

BIRDS.

Fringilla chloris. (Temminck.) The Greenfinch.
Very abundant.

Fringilla coccothraustes. (Temm.) The Hawfinch.
Has been shot in the winter months.

Fringilla carduelis. (Linn.) The Goldfinch—local name, Proud Tailor. Not so common as formerly.

Fringilla spinus. (Linn.) The Siskin.
A winter visitor.

Fringilla cannabina. (Linn.) The Linnet—local name, the Brown Linnet.

Fringilla linaria. (Linn.) The Lesser Redpole.
Rare.

Loxia pyrrhula. (Penn.) The Bullfinch.
Common.

Loxia curvirostra (Linn.) The Crossbill.
Sir John Crewe records many as killed from a flock in 1837-8.

FAMILY—STURNIDÆ.

Sturnus vulgaris. (Linn.) The Starling.
Very common. Is seen in the winter in large flocks, associating with Rooks.

FAMILY—CORVIDÆ.

Corvus corax. The Raven.
Once seen at Calke Park.

Corvus corone. (Linn.) The Carrion Crow.

 The ruthless band of the gamekeeper has made this a comparatively rare bird. Nests are, however, found every season in the large woods at Bretby and Ingleby.

Corvus cornix. (Linn.) The Hooded Crow.

 Has been shot at Calke Park, but is very rare.

Corvus frugilegus. (Linn.) The Rook.

 The large Rookeries in the neighbourhood cause this to be almost (if not quite) our commonest bird.

Corvus monedula. (Linn.) The Jackdaw.

 Very common. Builds at Foremark, and in the tower of the church.

Corvus pica. (Penn.) The Magpie.

Corvus glandarius. (Penn.) The Jay.

 Both the latter are common, the Magpie especially so. The Jay builds in Repton Shrubs.

TRIBE—SCANSORES.

FAMILY—PICIDÆ.

Picus viridis. (Linn.) The Green Woodpecker.

Picus major. (Linn.) The Great Spotted Woodpecker.

Picus minor. (Linn.) The Lesser Spotted Woodpecker.

Yunx torquilla. (Linn.) The Wryneck.

 All this family are rare, and but seldom breed with us; many instances, however, are on record of their occurrence in Calke Park.

FAMILY—CUCULIDÆ.

Cuculus canorus. (Linn.) The Cuckoo.

Very common. Many eggs are taken from the nests of the Reed Sparrow (*Salicaria arundinacea*). I once took an egg (which is now in my possession) from a Wagtail's nest, in the Cricket-field wall. The nest was built in a small hole, and I should not have believed it possible that a Cuckoo could have passed through the aperture, had I not observed the old Cuckoo flying from the nest.*

TRIBE—TENUIROSTRES.

FAMILY—CERTHIADÆ.

Certhia familiaris. (Linn.) The Tree Creeper.

Very common. Remains throughout the year.

Sylvia troglodytes. (Penn.) The Wren—local name, Jinty.

Sitta europæa. (Penn.) The Nuthatch.

Is often met with at Foremark. The eggs are but rarely taken.

TRIBE—FISSIROSTRES.

FAMILY—HALCIONIDÆ.

Alcedo ispida. (Linn.) The Kingfisher.

This beautiful bird is still of tolerably frequent occurrence, but I fear that each year, as it passes, sees a diminution of its numbers. A pair have for many

---

* Compare the statement of Le Valliant as to the manner in which this bird deposits her eggs. "Wood's Illustrated Natural History"—Birds—p. 572. S. A. P.

years built in the bank of a pond at Newton Park, where they are, and have been every year (with one exception), permitted to rear their young without molestation.

FAMILY—HIRUNDINIDÆ.

Hirundo rustica. (Linn.) The Swallow.

Hirundo urbica. (Linn.) The House Martin.

Hirundo riparia. (Linn.) The Sand Martin.

All very common.

Cypselus apus. (Illig.) The Swift—local name, the Squealer.

Very common. Builds under the eaves of the chancel of the church, and in the roofs of the old thatched houses of the town.

FAMILY—CAPRIMULGIDÆ.

Caprimulgus europœus. (Linn.) The Nightjar.

Is sometimes heard, but no instance is recorded of its building with us.

ORDER—RASORES.

FAMILY—COLUMBIDÆ.

Columba palumbus. (Linn.) The Ring Dove—local name, the Wood Pigeon.

Columba œnas. (Linn.) The Stock Dove.

By no means rare. I formerly took many nests in a season, especially in the neighbourhood of Egginton.

Columba livia. (Briss.) The Rock Dove.
  At Anchor Church, and Foremark; from the latter place I have the eggs. (W. G.)

Columba turtur. (Linn.) The Turtle Dove.
  Pair shot at Twyford. Mr. E. A. Brown has several times seen specimens in Drakelowe Park.

FAMILY—PHASIANIADÆ.

Phasianus colchicus. (Linn.) The Pheasant.
  It is owing to the excessive preservation of this and other game birds in the county that some of our more interesting birds are of so rare occurrence.

FAMILY—TETRAONIDÆ.

Lagopus scoticus. (Linn.) The Red Grouse.
  Has been seen occasionally, doubtless wanderers from the moors in the north of the county, or from Cannock Chase. One in an exhausted state was found many years ago (1865) on the road between Newton Solney and Burton.

Tetrao perdix. (Linn.) The Partridge.

Tetrao rufus. (Bewick.) The Red-legged Partridge.
  A bird of this species has been shot this season (1865), in the neighbourhood. It is, however, very rarely met with. 1881—Mr. E. A. Brown tells me that several have been killed in the last few years at Newton Solney.

Perdix coturnix. (Latham.) Common Quail.
  Sir John Crewe says it was formerly common at Swarkestone. Has been known to breed.

  (I have the eggs from Wain's Farm, Bretby. W. G.)

ORDER — GRALLATORES.
FAMILY—CHARADRIDÆ.

Charadrius pluvialis. (Linn.) The Golden Plover.
  Shot sometimes in the winter.

Charadrius hiaticulla. (Penn.) The Ringed Plover.
  Scarce on the Trent near Burton.

Charadrius morinellus. (Linn.) The Dotterel.
  Mr. Brown, in his History of Tutbury, records an occurrence of this species at Twyford.

Vanellus cristatus. (Flem.) The Lapwing—local name, the Peewit.
  Very common in the winter.

Hæmatopus ostralegus. (Linn.) The Oyster Catcher.
  A straggler of this species has been shot on the Trent, at Newton Solney.

FAMILY—ARDEIDÆ.

Ardea cinerea. (Linn.) The Common Heron—local name, the Hern.
  Builds occasionally at Anchor Church. I hear that a pair built there this year (1865).

Ardea purpurea. (Linn.) The Purple Heron.
  Shot at Newton Solney within last few years (1881).

Ardea stellaris. (Penn.) The Bittern.
  An old Reptonian shot a bird of this species, in immature plumage, in the winter of 1862-63, near the junction of the Dove and Trent, and I have a specimen killed on the Trent before the above date.

FAMILY—SCOLOPACIDÆ.

Totanus ochropus. (Temm.) The Green Sandpiper.
  Has occasionally been seen.

Totanus hypoleucus. (Temm.) The Sandpiper.
: Common on the Trent; occasionally builds.

Scolopax rusticola. (Linn) The Woodcock.
: Is not common.

Scolopax gallinago. (Linn.) The Full Snipe.
: Is said to breed with us, but I have no well-authenticated instance. It is common in winter.

Scolopax gallinula. (Linn.) The Jack Snipe.
: Not so common as the above.

Tringa variabilis. (Selby.) The Dunlin.
: Sir John Crewe states not uncommon in winter.

FAMILY—RALLIDÆ.

Crex pratensis. The Land Rail—local name, the Corncrake.
: This singular ventriloquist is to be heard every night in the summer.

Crex porzana. (Selby.) The Spotted Crake.
: One was killed some years ago near Willington, by flying against the telegraph wires.
: I have had one egg of this bird given to me from the Old Trent, and I have seen young ones. (W. G.)

Rallus aquaticus. (Linn.) The Water Rail.
: Rare.

Gallinula chloropus. (Lath.) The Moor or Water Hen.
: Very common.

FAMILY—LOBIPEDIDÆ.

Fulica atra. (Linn.) The Bald Coot.
: Not so common as formerly. Builds still on Repton Park Pool, where two broods have been hatched this year (1865). Hartshorne Pool was formerly a favourite "habitat."

ORDER—NATATORES.

FAMILY—ANATIDÆ.

Cygnus ferus. (Flem.) The Wild Swan.
> Has been occasionally shot.

Cygnus Bewickii. (Bewick.) Bewick's Swan.
> A specimen (almost the only one recorded in the country) was shot by Mr. C. Smallwood, an old Reptonian, at Newton, in the winter of 1863-4.

Cygnus olor. (Bois.) The Mute Swan.
> Too numerous, as they are sadly destructive to the spawn of fish.

Anser phœnicopus. (Bartlett.) Pink-footed Goose.
> Specimens killed at Newton Solney, 1877 and 1880.
> The Rev. John Wadham records one killed at Weston, in 1869-70.

Anser erythropus. (Fleming.) White-fronted Goose.
> Is rarely met with.

Anas tadorna. (Linn.) The Shieldrake.
> Two killed at Newton Solney, 1865.

Anas acuta. (Linn.) The Pintail.

Anas boschas. (Linn.) The Wild Duck. Breeds with us.

Anas crecca. (Linn.) The Teal. Not rare in winter.

Anas penelope. (Linn.) The Wigeon.
> Common on the Trent.

Anas nigra. (Penn.) Common Scoter.
> Sir John Crewe thinks tolerably plentiful on the Trent some winters.

Anas clypeata. (Penn.) The Shoveller.
> Shot by Rev. John Wadham at Barrow, 1863-4.

Anas Marila. The Scaup Duck.
> Shot at Weston, November 13, 1876.

Anas fusca. (Penn.) The Velvet Scoter.

This bird, so rarely met with inland, was killed by the Rev. J. Smith at Willington some years ago (1865).

Fuligula ferina. (Steph.) The Pochard.

Has been killed several times on the Trent in hard weather.

Fuligula cristata. (Steph.) The Tufted Duck.

Specimens killed on the Trent, at Newton Solney, 1879 and 1880, also Rev. John Wadham shot one at Weston, December 28, 1874.

Anas glacialis. (Penn.) The Long-tailed Duck.

A specimen killed near Twyford is now in the collection of Sir John Crewe, Bart.

Anas clangula. (Pennant.) Goldeneye.

A specimen recorded by Mr. E. A. Brown at Burton, in 1881; and this bird has been seen at Twyford frequently, one shot at Weston, December 16, 1874.

Mergus albellus. (Penn.) The Smew.

Killed by my brother (the late Mr. Francis Worthington) at the Hargate, in 1869.

FAMILY—COLYMBIDÆ.

Mergus Merganser. (Penn.) Goosander.

Newton Solney in 1879-80.

Podiceps cornutus. (Lath.) The Sclavonian Grebe.

Shot at Newton Solney, in 1860.

Podiceps rubricollis. (Lath) Red-necked Grebe.

A pair of these birds nested at Willington a few years ago; and the late Rev. F. M. Spilsbury took a couple of the eggs. (W. G.)

Podiceps minor. (Lath.) The Little Grebe.

Frequent on the Old Trent and Repton Park Pool.

I know of only one instance of its building with us, about 1859, on the millpond.

Colymbus glacialis. (Linn.) The Great Northern Diver.
Very rare. Shot on the Trent, at Newton Solney, in 1861-2.

Colymbus septentrionalis. (Linn.) Red-throated Diver.
Has been killed on Repton Park Pool.

Puffinus anglorum. (Temm.) The Manx Shearwater.
One killed at Newton Solney a few years ago, and in my possession. (W. G.)

Pelecanus Bassanus. . (Penn.) Common Gannet.
A specimen killed at Willington years ago.

FAMILY—LARIDÆ.

Sterna hirundo. (Linn.) The Common Tern.
Common in boisterous weather. Have been shot at Barrow.

Sterna arctica. (Temm.) The Arctic Tern.
Has occasionally been seen.

Larus tridactylus. (Linn.) The Kittiwake Gull.
I remember an instance of this bird occurring on the Old Trent, in a very hard winter. Specimen killed at Newton Solney, Jan., 1880.

Larus canus. (Linn.) The Common Gull.
Met with on the Trent.

Larus ridibundus. (Penn.) Black-headed Gull.
Several have occurred at Newton Solney.

N.B.—There are other specimens of Terns, Gulls, and Skuas which might be added to this list, but they are of such rare occurrence that I do not think it necessary to insert them.

A LIST

OF

LAND AND FRESHWATER SHELLS

FOUND

AT AND NEAR REPTON.

AQUATIC.

CLASS I.—CONCHIFERA OR BIVALVES.

Order.—LAMELLIBRANCHIATA.

Family I.—SPHÆRIIDÆ.

SPHÆRIUM.
    corneum. *Linn.* — Willington Canal, etc.
        var. flavescens. *Macgil.* — Bretby Park Ponds.
    rivicola. *Leach.* — Willington Canal.
    ovale. *Fer.* — River Trent—rare.

PISIDIUM.
    amnicum. *Mull.* — Old Trent, etc.
    fontinale. *Drap.* — Trent at Newton Solney.
        var. pulchellum. *Jenyns.* — River Trent.
    pusillum. *Gmelin.* — Willington Pits.

Family II.—UNIONIDÆ.

UNIO.
  tumidus. *Phil.* — Repton Park Pond and Willington Canal.
    var. radiata. — Repton Park Pond.
    var. ovalis. *Mont.* — Repton Park Pond.
  pictorum. *Lin.* * — Repton Park Pond and Willington Canal. *
    var. radiata. — Repton Park Pond—rare.
    var. curvirostris. *Norm.* — Repton Park Pond.
    var. latior. — Repton Park Pond and Willington Canal.

ANODONTA.
  cygnea. *Linn.* — Repton Park Pond and Willington Canal.
    var. Zellensis. *Gmelin.* — Repton Park Pond.
    var. radiata. *Mull.* — Bretby Mill Pond.
    var. rostrata. *Rossm.* — Repton Park Pond.
  * anatina. *Linn.* — Repton Park Pond.
    var. radiata. — Repton Park Pond.
    var. ventricosa. *Pf.* — Repton Park Pond.
    var. complanata. *Rossm.* — Repton Park Pond.

Family III.—DREISSENIDÆ.

DREISSENA.
  polymorpha. *Pallas.* — Willington Canal.

---

* Some few years ago, Repton Park Pond burst, when, in addition to the Unios and Anodons above given, several specimens were taken which scarcely come under the head of *named* varieties. It may also be further noted that several years ago a remarkable variety, which in "Brown's Conchology" is named *Anodonta contorta*, was found by a gentleman named Glover in some ditches near to the Old Trent, but this shell has not since been found.

CLASS II.—GASTEROPODA OR UNIVALVES.

Order I.—PECTINIBRANCHIATA.

Family I.—NERITIDÆ.

NERITINA.
  fluviatilis. *Linn.*   Willington Canal.

Family II.—PALUDINIDÆ.

PALUDINA.
  vivipara. *Linn.*   Willington Canal.

BYTHINIA.
  tentaculata. *Linn.*   Repton Top Mill-dam, and Trent.
    var. ventricosa. *Monke.*   Repton Top Mill-dam.
    var. albida.   Repton Top Mill-dam.
    Leachii. *Shepp.*   Willington Pits, etc.

Family III.—VALVATIDÆ.

VALVATA.
  piscinalis. *Mull.*   Trent, near Newton Solney.
  cristata. *Mull.*   Egginton, in Ditches.

Order II.—PULMONOBRANCHIATA.

Family—LIMNÆIDÆ.

PLANORBIS.

albus. *Mull.*	Pond at Spurs Bottoms, etc.
glaber. *Jeff.*	Pits at Willington.
spirorbis. *Mull.*	Old Trent, etc.
vortex. *Linn.*	Very common.
carinatus. *Mull.*	Common.
complanatus. *Linn.*	Common.
var. albida.	Repton Mill Top-dam.
corneus. *Linn.*	Old Trent—common.
contortus. *Linn.*	Old Trent, etc.

PHYSA.

hypnorum. *Linn.*	In ditches.
fortinalis. *Linn.*	River Trent—common.

LIMNÆA.

peregra. *Mull.*	Common.
var. intermedia. *Fer.*	Repton Park Pond.
auricularia. *Linn.*	Repton Park Pond.
stagnalis. *Linn.*	Old Trent—common.
var. fragilis. *Linn.*	Willington Pits.
palustris. *Mull.*	Old Trent.
var. elongata.	Old Trent.
var. tincta. *Jeff.*	Old Trent.
var. albida.	Old Trent.
truncatula. *Mull.*	Marshes close to Old Trent.

ANCYLUS.
    fluviatilis. *Mull.*   Repton Brook.
      var. albida.   Repton Brook.
    lacustris. *Linn.*   River Trent near Newton Solney.

      var. albida.   River Trent near Newton Solney.

TERRESTRIAL.

Family I.—LIMACIDÆ.

ARION.
    ater. *Linn.*   Everywhere.

LIMAX.
    gagates. *Drap.*   Hedges, roots of grass, etc.
    marginatus. *Mull.*   Under stones, among dead leaves, etc.
    flavus. *Linn.*   Cellars, wells, damp places.
    agrestis. *Linn.*   Fields, gardens, and woods.
    maximus. *Linn.*   Woods, under old logs, etc.

Family III.—HELICIDÆ.

SUCCINEA.
    putris. *Linn.*   Osier-beds—common.
    elegans. *Risso.*   Repton Rocks.

VITRINA.
    pellucida. *Mull.*   Among moss, dead leaves, woods, etc.

ZONITES.

cellarius. *Mull.*	Cellars, drains, woods, etc.
var. complanata.	Cellars, drains, woods, etc.
var. albida.	Cellars, drains, woods, etc.
var. compacta.	Cellars, drains, woods, etc.
alliarius. *Miller.*	Under stones, foot of old walls, etc.
glaber. *Studer.*	Under stones, foot of old walls, etc.
nitidulus. *Drap.*	Under stones, dead leaves.
purus. *Alder.*	In moss, dead leaves, etc.
radiatulus. *Alder.*	Foot of old walls, in moss.
nitidus. *Mull.*	Roots of grass, moist places.
excavatus. *Bean.*	Robin's Wood, under fallen wood, etc.
var. vitrina. *Fer.*	Robin's Wood, under fallen wood, etc.
crystallinus. *Mull.*	Under stones, decayed wood, etc.
fulvus. *Mull.*	Robin's Wood.
var. Mortoni. *Jeffr.*	Robin's Wood.

HELIX.

aculeata. *Mull.*	Robin's Wood.
aspersa. *Mull.*	In gardens, and Gravel Pit Repton.
nemoralis. *Linn.*	Hedge banks—common.
var. hortensis. *Mull.*	Hedge banks—common.
var. hybrida. *Poiret.*	Burton Road.
var. minor.	Burton Road.

arbustorum. *Linn.*   Hedge banks and osier-beds.
   var. albida.   Milton and Foremark Road.
   var. flavescens.   Milton and Foremark Road.
concinna. *Jeffr.*   Under stones, among nettles, etc.
hispida. *Linn.*   Under stones, logs of wood, etc.
   var. subrufa.   Near Bull's, Meadow Farm.
virgata. *Da Costa.*   Near Ticknall.
caperata. *Mont.*   Ticknall Lime Quarry.
rotundata. *Mull.*   Under stones—common.
pygmæa. *Drap.*   Robin's Wood.
pulchella. *Mull.*   Roots of grass, foot of old walls.
   var. costata. *Mull.*   Roots of grass, foot of old walls.

BULIMUS.
obscurus. *Mull.*   Old walls, etc.

PUPA.
umbilicata. *Drap.*   Under trees, etc.
   var. edentula.   Robin's Wood.
marginata. *Drap.*   Robin's Wood.

VERTIGO.
pygmæa. *Drap.*   Under stones and flood refuse.
edentula. *Drap.*   Bretby Wood.

BALIA.
perversa. *Linn.*   Under bark of willows, Trent Meadows.

CLAUSILIA.

   rugosa. *Drap.*   On old walls, etc.
   laminata. *Mont.*   In woods at Shobnall.

COCHLICOPA.

   lubrica. *Mull.*   Under logs of wood, dead leaves, etc.

      var. lubricoides. *Fer.*   Under logs of wood, dead leaves, etc.

ACHATINA.

   acicula. *Mull.*   Roots of bushes and grass.

FAMILY IV.—CARYCHIIDÆ.

CARYCHIUM.

   minimum. *Mull.*   Under stones, roots of grass, moss, etc.

A LIST OF COLEOPTERA

OF

REPTON AND NEIGHBOURHOOD.

CICINDELIDÆ.

CICINDELA.
    campestris. L. Rare.

CARABIDÆ.

NOTIOPHILUS.
    aquaticus. L. Common.
    palustris. Duft. Common.
    biguttatus. F. Common.
    substriatus. Wat. Rare.

ELAPHRUS.
    riparius. L. Common.
    cupreus. Duft. Common.
    uliginosus. F. Robin's Wood.

BLETHISA.
    multipunctata. L. Robin's Wood.

CYCHRUS.
    rostratus. L. Occasionally.

CARABUS.
    granulatus. L. Common.
    monilis. F. Common.
    catenulatus. Scop. Not rare.
    nemoralis. Müll. Common.
    violaceus. L. Common.

CALOSOMA.
    * inquisitor. L. Seal Wood.

NEBRIA.
    brevicollis. F. Very common.

PELOPHILA.
    borealis. Pk. Robin's Wood. Very rare.

LEISTUS.
    spinibarbis. F. Common.
    fulvibarbis. Dj. Common.
    ferrugineus. L. Common.
    rufescens. F. Common.

CLIVINA.
    fossor. L. Common.
    collaris. Hbst. Not rare by side of Trent and Dove.

DYSCHIRIUS.
    globosus. Hbst. Not common.

DEMETRIAS.
    atricapillus. L. Common.

---

* Seal Wood is beyond the Repton district, properly so called, but being a large and handsome beetle, it is deemed advisable to record it.

DROMIUS.
    linearis. Ol. Common.
    meridionalis. Dj. Occasionally.
    agilis. F. Common.
    quadrimaculatus. L. Common.
    quadrinotatus. Pz. Common.
    melanocephalus. Dj. Common.

METABLETUS.
    foveola. Gyll. Occasionally.

LEBIA.
    chlorocephala. E. H. Once taken.

LORICERA.
    pilicornis. F. Very common.

CHLÆNIUS.
    vestitus. Pk. Rare.
    nigricornis. F. Not common.
   * holosericeus. F.

OODES.
    helopioides. Not common.

BADISTER.
    bipustulatus. F. Common.
    sodalis. Duft. Rare—in woods.

BROSCUS.
    cephalotes. L. Once in Repton street.

SPHODRUS.
    leucopthalmus. L. Once in Repton.

---

* What I believe to be this insect I have once taken here.

PRISTONYCHUS.
subcyaneus. Ill. Occasionally in out-houses.

CALATHUS.
cisteloides. Pz. Very common.
melanocephalus. L. Very common.
piceus. Marsh. Not rare at Foremark.

TAPHRIA.
nivalis. Pz. Not common—in woods.

ANCHOMENUS.
junceus. Scop. Very common in woods.
prasinus. Thunb. Common.
albipes. F. Common.
marginatus. L. Common.
parumpunctatus. F. Common.
viduus. Pz. In osier-beds. Rare.
v. mœstus. Duft. In wet places. Not rare.
versutus. Gyll. Not common.
atratus. Duft. Not common.
micans. Nic. Common in wet places.
scitulus. Dj. Very rare.
piceus. L. Common in wet places.
gracilis. Gyll. Rare.
fuliginosus. Pz. Very common.
Thoreyi. Dj. Rare.
puellus. Dj. Not rare—in osier-beds.

OLISTHOPUS.
rotundatus. Pk. Under stones.

STOMIS.
pumicatus. Pz. In flood refuse.

PLATYDERUS.
    ruficollis. Marsh. Not common.

PTEROSTICHUS.
    cupreus. L. Common.
    versicolor. Sturm. Not uncommon.
    picimanus. Duft. Very rare.
    vernalis. Pz. Common.
    niger. Schal. Common.
    melanarius. Ill. Common.
    anthracinus. Ill. Rare.
    nigrita. F. Very Common.
    gracilis. Dj. Rare—at Egginton.
    minor. Gill. Rare—at Crewe's Pond.
    strenuus. Pz. Common.
    diligens. Sturm. Not uncommon at Robin's Wood.
    oblongo-punctatus. F. Robin's Wood.
    madidus. F. Very common.
    striola. F. Common.

AMARA.
    apricaria. Pk. Common.
    consularis. Duft. Very rare.
    spinipes. L. Not common.
    bifrons. Gyll. Once in Repton.
    familiaris. Duft. Common.
    acuminata. Pk. Not common.
    trivialis. Gyll. Common.
    lunicollis. Schiod. Not common.
    communis. Pz. Very common.
    ovata. F. Not common.

similata. Gyll.   Not common.
strenua. Zim.   Once in Repton street.
plebeia. Gyll.   Very common.

HARPALUS.
puncticollis. Pk.   Not common.
rufilabris. F.   Local—Askew Hill.
ruficornis. F.   Very Common.
consentaneus. Dj.   Scarce.
Proteus. Pk.   Very common.
latus. L.   Scarce.
tardus. Pz.   Scarce.

ACUPALPUS.
meridianus. L.   Not rare.
exiguus. Dj.
v. luridus. Dj.   Near Burton.

BRADYCELLUS.
distinctus. Dj.   Robin's Wood.
verbasci. Duft.   Robin's Wood.
harpalinus. Dj.   Robin's Wood.

PATROBUS.
excavatus. Pk.   Common.

TRECHUS.
discus. F.   By Trent and Dove.   Rare.
micros. Hbst.   By Trent and Dove.   Not common.
rubens. F.   Egginton.   Very rare.
minutus. F.   Common.
obtusus. Er.   Common.
secalis. Pk.   Woods.   Not rare.

BEMBIDIUIM.

rufescens. Guer. Not rare.
quinquestriatum. Gyll. Rare. On walls.
obtusum. Sturm. Common.
biguttatum. F. Common.
æneum. Germ. Rare, by the Trent.
guttula. F. Common.
Mannerheimi. Sahl. Not rare in Woods.
quadrimaculatum. L. Common in sand pits.
quadripustulatum. Dj. Once by the Trent.
quadriguttatum. F. Common in sand pits.
articulatum. Pz. Not common.
gilvipes. Sturm. Not common, in flood refuse.
lampros. Hbst. Common.
decorum. Pz. By the Dove at Egginton.
monticola. Sturm. By the Dove at Egginton.
Stephensi. Dj. In sand pits. Not common.
brunnipes. Sturm. Common in sand pits.
tibiale. Duft. Common by the Dove at Egginton.
atrocæruleum. Steph. By the Dove at Egginton.
femoratum. Sturm. Not uncommon; at Bull's in the Meadow.
bruxellense. Wesm. Robin's Wood.
littorale. Ol. Very common.
fluviatile. Dj. By the Trent and Dove.
flammulatum. Clair. By the Trent and Dove. Very common.
punctulatum. Drap. By the Trent and Dove. Not uncommon.

TACHYPUS.

flavipes. L. Not uncommon about Repton.

DYTISCIDÆ.

HALIPLUS.
    obliquus. F. Findern.
    mucronatus. Steph. In a ditch at Burton.
    fulvus. F. Ponds and ditches. Not common.
    cinereus. Aubé. Repton brook and ditches at Twyford. Rare.
    fluviatilis. Aubé. Common in the river.
    ruficollis. De G. Common in ditches.
    lineatocollis. Marsh. Common in ditches.

BRYCHIUS.
    elevatus. Pz. Repton Brook, and Dove at Egginton.

PELOBIUS.
    Hermanni. F. Once at Twyford.

HYDROPORUS.
    reticulatus. F. Common in ditches.
    quinquelineatus. Zett. Very rare.
    inæqualis. F. Common.
    pictus. F. Common.
    duodecimpustulatus. Ol. Dove at Egginton.
    dorsalis. F. Twyford. Not common.
    memnonius. Nic. Twyford.
    erythrocephalus. L. Very common.
    marginatus. Duft. Twyford and Egginton. Very rare.
    lituratus. F. Not uncommon in ponds in woods.
    planus. F. Not uncommon in ditches.
    melanocephalus. Steph. Very common.
    nigrita. F. Common.

tristis. Pk. Rare.
palustris. L. Very common.
angustatus. Sturm. Rare.
lineatus. F. Very common.

NOTERUS.
clavicornis. De G. Old Trent. Not common.
sparsus. Marsh. Very common.

LACCOPHILUS.
minutus. L. Common.
hyalinus. De G. Common.

COLYMBETES.
fuscus. L. Common.
pulverosus. Steph. Willington Pits. Not common.
exoletus. Forst. Willington Pits. Not common.

ILYBIUS.
fenestratus. F. Common.
fuliginosus. F. Common.
ater. De G. Not uncommon.
obscurus. Marsh. Rare.

AGABUS.
bipustulatus. L. Very common.
chalconotus. Pz. Not common.
Sturmi. Schon. Common.
paludosus. F. Common.
guttatus. Pk. Not common.
didymus. Ol. Not common.
nebulosus. Forst. Willington Pits.
maculatus. L. Repton Brook.

DYTISCUS.
    marginalis. L. Occasionally in ponds.

ACILIUS.
    sulcatus. L. Common.

GYRINIDÆ.

GYRINUS.
    natator. Scop. Common.
    marinus. Gyll. Common.

ORECTOCHILUS.
    villosus. Mull. By Dove and Trent.

HYDROPHILIDÆ.

HYDROBIUS.
    fuscipes. L. Common in ditches.

HELOCHARES.
    lividus. Forst. Ditches.
    punctatus. Sharp. Ditches.

PHILHYDRUS.
    nigricans. Zett. Ponds and ditches.
    melanocephalus. Ol. Ponds and ditches.
    marginellus. F. Willington Pits.
    suturalis. Sharp. Willington Pits.

ENOCHRUS.
    bicolor. Pk. Pond at Findern.

ANACÆNA.
>limbata. F. Willington Pits.
>variabilis. Sharp. Willington Pits.

LACCOBIUS.
>minutus. L. Very common.
>nigriceps. Th. Willington Pits.

LIMNOBIUS.
>truncatellus. Thunb. Very common.
>papposus. Muls. Not common.

CHÆTARTHRIA.
>seminulum. Pk. Willington Pits.

HELOPHORUS.
>rugosus. Ol. Not common.
>nubilus. F. Not common.
>aquaticus. L. Common.
>Mulsanti. Rye. Old Trent.
>griseus. Hbst. Common.
>granularis. L. Common.
>æneipennis. Th. Old Trent.

HYDROCHUS.
>elongatus. Schall. Pond at Twyford. Rare.
>angustatus. Germ. Ponds. Scarce.

OCTHEBIUS.
>exsculptus. Germ. Egginton. Very rare.

HYDRÆNA.
>riparia. Kug. Old Trent.
>nigrita. Germ. Old Trent.
>gracilis. Germ. Old Trent.

CYCLONOTUM.
orbiculare. F. Willington Pits.

SPHÆRIDIUM.
scarabæoides. L. Common.
bipustulatum. F. Not common.
marginatum. F. Near Burton.

CERCYON.
obsoletus. Gyll. Very rare.
hæmorrhoidalis. F. Common.
hæmorrhous. Gyll. Common.
lateralis. Marsh. Flood refuse.
unipunctatus. L. In cucumber beds.
quisquilius. L. In cucumber beds.
melanocephalus. L. Common.
terminatus. Marsh. Cucumber beds.
pygmæus. Ill. Cucumber beds.
nigriceps. Marsh. Cucumber beds.
minutus. F. Cucumber beds.
lugubris. Pk. Cucumber beds.
granarius. Er. Cucumber beds.
analis. Pk. Cucumber beds.

MEGASTERNUM.
boletophagum. Marsh. Common.

CRYPTOPLEURUM.
atomarium. F. Common.

STAPHYLINIDÆ.

AUTALIA.
    impressa. Ol. Under garden refuse and in cucumber beds.
    rivularis. Gr. Common.

FALAGRIA.
    sulcata. Pk. In cucumber beds.
    obscura. Gr. In cucumber beds.

BOLITOCHARA.
    bella. Märk. In a fungus in Twyford Road.

OCALEA.
    latipennis. Sharp. In fungi at Bretby Wood.
    castanea. Er. In fungi at Bretby Wood.
    badia. Er. In fungi at Bretby Wood.

ISCHNOGLOSSA.
    corticina. Er. Under bark of trees.
    rufopicea. Kr. Under bark of trees.

ALEOCHARA.
    ruficornis. Gr. Very rare. Findern Gravel-pit.
    fuscipes. F. Common.
    bipunctata. Ol. Rare.
    brevipennis. Gr. Rare.
    lanuginosa. Gr. Common.
    mœsta. Gr. About Repton.
    nitida. Gr. Under garden refuse.
    V. bilineata. Gyll. Under garden refuse.

MYRMEDONIA.
    canaliculata. F. Common under stones.

CHILOPORA.
longitarsis. Steph. In Bull's meadow, and flood refuse.

TACHYUSA.
constricta. Er. Newton Solney.
flavitarsis. Sahl. By the Trent and Dove.
atra. Gr. Not uncommon in flood refuse.

OXYPODA.
lividipennis. Mann. Garden refuse.
vittata. Märk. Garden refuse.
opaca. Gr. Garden refuse.
longiuscula. Gr. Common in flood refuse.
alternans. Gr. Robin's Wood.
incrassata. Muls. Rare.

HOMALOTA.
gregaria. Er. Flood refuse.
vicina. Steph. Flood refuse.
graminicola. Gr. Flood refuse.
æquata. Er. Near Repton.
angustula. Gyll. Near Repton.
debilis. Er. Near Repton.
circellaris. Gr. Common under stones.
cuspidata. Er. Under bark of trees. Rare.
depressa. Gyll. Repton.
trinotata. Kr. Repton.
triangulum. Kr. Repton.
fungicola. Th. Repton.
boletobia. Th. Repton.
coriaria. Kr. Repton.
palustris. Kies. Near Repton.

cinnamomea. Gr. Near Repton.
hospita. Mark. Near Repton.
sericea. Muls. Near Repton.
marcida. Er. Near Repton.
longicornis. Gr. Near Repton.
melanaria. Sahl. Near Repton.
aterrima. Gr. Near Repton.
pygmæa. Gr. Near Repton.
fusca. Sahl. Near Repton.
fungi. Gr. Very common.

PHLŒOPORA.
reptans. Gr. Under bark of trees.
corticalis. Gr. Under bark of trees.

HYGRONOMA.
dimidiata. Gr. Bretby Wood.

OLIGOTA.
inflata. Mann. Foot of old haystacks.
pusillima. Gr. Foot of old haystacks. Rare.
ruficornis. Sharp. Foot of old haystacks. Common.

ENCEPHALUS.
complicans. Steph. In flood refuse—Mr. Fowler.

GYROPHÆNA.
gentilis. Er. In fungi. Repton Shrubs.
affinis. Mann. In fungi. Repton Shrubs.
nana. Pk. In fungi. Repton Shrubs.
congrua. Er. In fungi. Repton Shrubs.
lævipennis. Kr. In fungi. Repton Shrubs.
minima. Er. In fungi. Repton Shrubs.
manca. Er. In fungi at Repton and Stenson.

AGARICOCHARA.
lævicollis. In fungi at Repton. Very rare.

HYPOCYPTUS.
longicornis. Pk. Common.

CONURUS.
littoreus. L. Haystack refuse.
pubescens. Gr. Haystack refuse.
lividus. Er. By ditch at the tanyard.

TACHYPORUS.
obtusus. L. In moss.
chrysomelinus. L. In moss.
pallidus. Sharp. In moss.
humerosus. Er. In moss.
hypnorum. F. In moss.
tersus. Er. In moss.
transversalis. Gr. Repton Rocks.
brunneus. L. In moss.

LAMPRINUS.
saginatus. Gr. Flood refuse at Egginton. Very rare.

HABROCERUS.
capillaricornis. Gr. Hotbeds.

CILEA.
silphoides. L. Hotbeds.

TACHINUS.
humeralis. Gr. Bretby and Robin's Wood.
pallipes. Gr. Egginton.
rufipes. De G. Common.

scapularis. Steph. Robin's Wood.
bipustulatus. F. Robin's Wood.
subterraneus. L. Under garden refuse.
laticollis. Gr. In flood refuse.
marginellus. F. In flood refuse.
collaris. Gr. In flood refuse.
elongatus. Gyll. Seal Wood, by Mr. Harris.

MEGACRONUS.
inclinans. Gr. Bretby Wood—Mr. Fowler.
cingulatus. Mann. Bretby and Robin's Wood.
analis. Pk. Bretby and Robin's Wood.

BOLITOBIUS.
atricapillus. F. In fungi in woods.
trinotatus. Er. In fungi in woods.
exoletus. Er. In fungi in woods.
pygmæus. F. In fungi in woods.

MYCETOPORUS.
lucidus. Er. Rare. Near Repton.
punctus. Gyll. In flood refuse.
splendens. Marsh. In flood refuse.
longulus. Mann. In flood refuse.
lepidus. Gr. In flood refuse.
clavicornis. Steph. In garden and flood refuse.
splendidus. Gr. In garden and flood refuse.
longicornis. Kr. In garden and flood refuse.

HETEROTHOPS.
prævia. Er. Old leaves and grass—Mr. Harris.

QUEDIUS.
fulgidus. Gr. Common.

cruentus. Ol.   Robin's Wood.
impressus. Pz.   Hotbeds.
molochinus. Gr.   Under refuse.
tristis. Gr.   Under refuse.
fuliginosus. Gr.   Under refuse.
picipes. Mann.   Under refuse.
peltatus. Er.   Flood refuse.
umbrinus. Er.   Flood refuse.
maurorufus. Gr.   Flood refuse.
suturalis. Kies.   Flood refuse.
rufipes. Gr.   Flood refuse.
attenuatus. Gyll.   Flood refuse.
fulvicollis. Steph.   Flood refuse.
boops. Gr.   Flood refuse.

CREOPHILUS.
maxillosus. L.   Common.

LEISTOTROPHUS.
nebulosus. F.   Bretby and Robin's Wood.
murinus. L.   Bretby and Robin's Wood.

STAPHYLINUS.
pubescens. De G.   Common.
erythropterus. L.   Near Repton.

OCYPUS.
olens. Müll.   Repton.
brunnipes. F.   Common.
cupreus. Rossi.   Common.

PHILONTHUS.
splendens. F.   Parson's Hills and Foremark.
intermedius. Boisd.   Tanyard and Foremark.

laminatus. Creutz. Tanyard and Foremark.
succicola. Th. Foremark.
æneus. Rossi. Common.
addendus. Sharp. Foremark.
carbonarius. Gyll. Foremark.
decorus. Gr. Common.
politus. F. Common.
umbratilis. Gr. In garden refuse.
marginatus. F. In woods. Common.
varius. Gyll. In garden refuse.
albipes. Gr. In hotbeds
sordidus. Gr. In hotbeds.
fimetarius. Gr. In hotbeds.
cephalotes. Gr. In garden refuse.
ebeninus. Er. In hotbeds.
sanguinolentus. Gr. Under refuse.
bipustulatus. Pz. Under refuse.
scybalarius. Nord. In garden refuse.
varians. Pk. Common.
agilis. Gr. In garden refuse.
debilis. Gr. In garden refuse.
discoideus. Gr. Hotbeds.
ventralis. Gr. Hotbeds.
thermarum. Aub. Hotbeds—Mr. Fowler.
trossulus. Nord. In flood refuse.
nigritulus. Gr. In flood refuse.
puella. Nord. In garden refuse.
procerulus. Gr. In hotbeds.

XANTHOLINUS.
glabratus. Gr. Tanyard, etc.
punctulatus. Pk. Tanyard, etc.

ochraceus. Gyll. Common.
longiventris. Heer. Common.
linearis. Ol. Common.

LEPTACINUS.
parumpunctatus. Gyll. Foremark.
batychrus. Gyll. Common.
linearis. Gr. Common.

BAPTOLINUS.
alternans. Pk. Robin's Wood.

OTHIUS.
fulvipennis. F. Common.
punctipennis. Lac. Foremark Bottoms.
melanocephalus. Gr. Common.

LATHROBIUM.
brunnipes. F. Common.
boreale. Hoch. Flood refuse.
elongatum. L. Common.
fulvipenne. Gr. Common.
multipunctum. Gr. Willington Pits.
terminatum. Gr. Willington Pits.
longulum. Gr. Flood refuse.

ACHENIUM.
humile. Nic. Flood refuse.

CRYPTOBIUM.
fracticorne. Pk. Willington Pits.

STILICUS.
rufipes. Gr. Common.
affinis. Er. In haystack rubbish.
orbiculatus. Pk. In haystack rubbish.

LITHOCHARIS.
    obsoleta. Nord. Hotbed near Burton—Mr. Harris.
    melanocephala. F. Haystack refuse.
    propinqua. Bris. Haystack refuse.

SUNIUS.
    angustatus. Pk. Haystack refuse.

PÆDERUS.
    littoralis. Gr. Flood refuse, but rare here.

DIANOUS.
    cærulescens. Gyll. By Milton Brook.

STENUS.
    bipunctatus. Er. By the Dove at Egginton.
    guttula. Müll. By the Dove at Egginton.
    bimaculatus. Er. Common.
    Juno. F. Common.
    incrassatus. Er. In Bull's Meadow. Scarce.
    foveiventris. Fair. Common.
    cinerascens. Er. In Bull's Meadow. Rare.
    atratulus. Er. Bull's Meadow.
    melanopus. Marsh. Bull's Meadow. Scarce.
    canaliculatus. Gyll. Bull's Meadow. Scarce.
    æmulus. Er. Bull's Meadow. Scarce.
    pusillus. Steph. Common.
    speculator. Lac. Common.
    providus. Er. Not uncommon.
    carbonarius. Gyll. Tanyard. Scarce.
    argus. Gr. Scarce.
    circularis. Gr. Flood refuse. Scarce.
    declaratus. Er. Common.

crassiventris. Th.   Garden refuse.
unicolor. Er.   Common.
binotatus. Ljun.   Not uncommon.
pubescens. Steph.   Not uncommon at Egginton.
pallitarsis. Steph.   Not uncommon at Egginton.
bifoveolatus. Gyll.   Not common.
brevicornis. Th.   Not common.
picipennis. Er.   Not common.
rusticus. Er.   Very common.
tempestivus. Er.   Scarce.
subæneus. Er.   Scarce.
impressipennis. Duv.   Robin's Wood.
impressus. Germ.   Near Repton.
annulatus. Crotch.   Near Repton.
Erichsoni. Rye.   Not uncommon.
pallipes. Gr.   In moss, etc.
flavipes. Steph.   By sweeping—Bretby Wood.
cicindeloides. Gr.   Common.
similis. Hbst.   Common.
tarsalis. Ljun.   Common.
paganus. Er.   Common.

OXYPORUS.
rufus. L.   In the gills of fungi.   Robin's Wood.

BLEDIUS.
subterraneus. Er.   Newton Solney.
fracticornis. Pk.   Scarce.

PLATYSTETHUS.
arenarius. Fourc.   Common.
cornutus. Gr.   Common.

OXYTELUS.

    rugosus. F. Common.
    laqueatus. Marsh. Not uncommon.
    sculptus. Gr. Common.
    sculpturatus. Gr. Common.
    inustus. Gr. About Repton.
    nitidulus. Gr. Not uncommon.
    complanatus. Er. Common.
    tetracarinatus. Block. Common.
    speculifrous. Kr. Very rare.

HAPLODERUS.

    cælatus. Gr. Flood refuse. Not common.

TROGOPHLŒUS.

    arcuatus. Steph. Flood refuse.
    bilineatus. Steph. Flood refuse.
    Erichsoni. Sharp. Flood refuse.
    elongatulus. Er. Flood refuse.
    fuliginosus. Gr. Flood refuse.
    corticinus. Gr. Flood refuse.
    pusillus. Gr. Flood refuse.

SYNTOMIUM.

    æneum. Müll. Foremark. Not common.

COPROPHILUS.

    striatulus. F. Flood refuse. Not common.

DELEASTER.

    dichrous. Gr. Repton. Very rare.

LESTEVA.

longælitrata. Gœze. Common.
pubescens. Mann. Common.
punctata. Er. Common.

OLOPHRUM.

piceum. Gyll. Common in woods.

LATHRIMÆUM.

atrocephalum. Gyll. Under dead leaves, in woods.
unicolor. Steph. Under dead leaves, in woods.

HOMALIUM.

rivulare. Pk. Repton.
fossulatum. Er. Repton.
cæsum. Gr. Robin's Wood.
oxyacanthæ. Gr. Robin's Wood.
pusillum. Gr. Robin's Wood.
deplanatum. Gyll. Repton.
concinnum. Marsh. Repton.
vile. Er. Robin's Wood.
florale. Pk. Robin's Wood.
iopterum. Steph. Robin's Wood.
pygmæum. Gyll. Robin's Wood.

EUSPHALERUM.

primulæ. Steph. Repton Shrubs.

ANTHOBIUM.

opthalmicum. Pk. Common in flowers.
torquatum. Marsh. Common in flowers.

PROTEINUS.
    brevicollis. Er. Garden refuse.
    brachypterus. F. Garden refuse.

MEGARTHRUS.
    depressus. Pk. Garden refuse.
    affinis. Mill. Garden refuse.
    sinuatocollis. Lac. Garden refuse.
    denticollis. Beck. Garden refuse.

PHLŒOBIUM.
    clypeatum. Müll. Haystack refuse.

PROGNATHA.
    quadricorne. Kirb. Under bark of fallen trees.

MICROPEPLUS.
    porcatus. Pk. Rare. Mr. Harris.
    staphylinoides. Marsh. Hotbeds.
    Margaritæ. Duv. Hotbeds.

PSELAPHIDÆ.

BRYAXIS.
    fossulata. Reich. In moss.
    hæmatica. Reich. In moss.
    juncorum. Leach. In moss.

PSELAPHUS.
    Heisei. Hbst. Willington pits, in moss

TYCHUS.
    niger. Pk. Flood refuse. Very common.

BYTHINUS.
    puncticollis. Den. In moss. Not common.
    bulbifer. Reich. Common.
    Curtisi. Leach. Osier-bed at tanyard. Rare.
    securiger. Reich. In moss. Not common.
    Burrelli. Den. In moss. Not common.

EUPLECTUS.
    punctatus. Muls. Very rare. Bretby Wood.
    signatus. Reich. In hotbeds. Common.
    sanguineus. Aub. In hotbeds. Common.
    Karsteni. Reich. In hotbeds.
    minutissimus. Aub. Flood refuse.

SCYDMÆNIDÆ.

EUMICRUS.
    tarsatus. Müll. Common in hotbeds.

SCYDMÆNUS.
    scutellaris. Müll. In moss.
    collaris. Müll. In moss.
    angulatus. Müll. Very rare.
    elongatulus. Müll. Scarce.
    Sparshalli. Den. Loscoe and near Burton.
    fimetarius. Th. Common in hotbeds.

EUTHIA.
    Schaumi. Keis. Repton. Very rare.
    scydmænoides. Steph. In flood refuse and hotbeds.

CEPHENNIUM.
　thoracicum. Müll. Rare in moss.

SCAPHIDIIDÆ.

SCAPHISOMA.
　agaricinum. Ol. In fungi.
　boleti. Pz. In fungi.

TRICHOPTERYGIDÆ.

PTERYX.
　suturalis. Heer. Repton Shrubs—Mr. Fowler.

TRICHOPTERYX.
　thoracica. Gill. Flood refuse—Mr. Fowler.
　atomaria. De G. Common—Mr. Fowler.
　cantiana. Matth. Robin's Wood—Mr. Fowler.
　Lætitiæ. Matth. Robin's Wood—Mr. Fowler.
　longula. Matth. Hotbeds—Mr. Fowler.
　anthracina. Matth. Hotbeds—Mr. Fowler.
　fascicularis. Hbst. Robin's Wood—Mr. Fowler.
　lata. Mots. Common—Mr. Fowler.
　grandicollis. Mann. Very common in flood refuse
　　—Mr. Fowler.
　sericans. Heer. Hotbeds—Mr. Fowler.
　bovina. Mots. Repton Shrubs (?)—Mr. Fowler.
　Montandoni. All. Hotbeds—Mr. Fowler.

PTILIUM.
　Spencei. All. Repton shrubs. Mr. Fowler.

PTENIDIUM.
    pusillum. Gyll. Common.
    apicale. Er. Common.

CORYLOPHIDÆ.

ORTHOPERUS.
    atomus. Gyll. Robin's Wood.

SERICODORUS.
    lateralis. Gyll. Robin's Wood.

CLAMBIDÆ.

CALYPTOMERUS.
    dubius. Marsh. Robin's Wood.

CLAMBUS.
    pubescens. Redt. Hotbeds.
    armadillo. De G. Flood refuse.

ANISOTOMIDÆ.

AGATHIDIUM.
    nigripenne. F. Bretby Wood. Mr. Fowler.
    atrum. Pk. Robin's Wood.
    seminulum. L. Robin's Wood.
    varians. Beck. Robin's Wood.

AMPHICYLLIS.
    globus. F. Robin's Wood. Rare.

LIODES.
    humeralis. F. Common under dead sticks. Robin's Wood.

COLENIS.
    dentipes. Gyll. Bretby Wood.

ANISOTOMA.
    grandis. Fair. Bretby Wood. Mr. Fowler.
    dubia. Kug. Bretby Wood.
    calcarata. Er. Bretby Wood.
    badia. Sturm. Bretby Wood.
    litura. Steph. Near Burton. Mr. Harris.

SILPHIDÆ.

COLON.
    brunneum. Lat. Bull's, in the meadow.

CHOLEVA.
    angustata. F. Near Repton.
    Sturmi. Bris. Near Repton.
    spadicea. Sturm. Near Repton.
    agilis. Ill. Near Repton.
    fusca. Pz. Near Repton.
    nigricans. Spence. Near Repton.
    morio. F. Near Repton.
    nigrita. Er. Near Repton.
    tristis. Pz. Near Repton.
    grandicollis. Er. Near Repton.
    Kirbyi. Spence. Near Repton.
    chrysomeloides. Pz. Near Repton.
    Watsoni. Spence. Bretby Wood.
    fumata. Spence. Bretby Wood.
    velox. Spence. Bretby Wood.

Wilkini. Spence. Bretby Wood.
anisotomoides. Spence. Bretby Wood.
sericea. Pz. Common.

NECROPHORUS.
humator. F. Robin's Wood.
ruspator. Er. Robin's Wood.
mortuorum. F. Robin's Wood.
vespillo. L. Robin's Wood.

SILPHA.
thoracica. L. Foremark.
rugosa. L. Common.
dispar. Hbst. Very rare.
sinuata. F. Common.
nigrita. Er. Not common.
tristis. Ill. Not common.
lævigata. F. Not common.
atrata. L. Common.

HISTERIDÆ.

HISTER.
cadaverinus. E. H. Common.
succicola. Th. Foremark.
unicolor. L. Foremark.
merdarius. E. H. Foremark.
neglectus. Germ. Foremark.
carbonarius. E. H. Foremark.
12-striatus. Schr. Cucumber beds.
bimaculatus. L. Cucumber beds.

CARCINOPS.
minima. Aub. Flood refuse.

GNATHONCUS.
    rotundatus. Ill. Cucumber beds. Mr. Harris.

SAPRINUS.
    nitidulus. Pk. Common.

ONTHOPHILUS.
    striatus. F. Tanyard.

ABRÆUS.
    globosus. E. H. In fungi. Twyford Road.

ACRITUS.
    minutus. Pk. Hotbeds.
    nigricornis. E. H. In fungi. Very rare.

PHALACRIDÆ.

PHALACRUS.
    corruscus. Pk. Near Repton.

OLIBRUS.
    corticalis. Pz. Common.
    æneus. F. Twyford.
    consimilis. Marsh. Common.

NITIDULIDÆ.

CERCUS.
    pedicularius. L. Osier-beds.
    bipustulatus. Pk. Osier-beds.

BRACHYPTERUS.
    gravidus. Ill. Twyford.
    pubescens. Er. Common.
    urticæ. F. Common.

EPURÆA.
    æstiva. L. Common.
    melina. Er. Bretby Wood.
    deleta. Er. Common.
    obsoleta. F. Robin's Wood.
    florea. Er. Robin's Wood.
    melanocephala. Marsh. Foremark. Mr. Fowler.
    limbata. F. In fungi.

NITIDULA.
    bipustulata. L. Robin's Wood.

SORONIA.
    punctatissima. Ill. Not common.
    grisea. L. Common.

OMOSITA.
    colon. L. Not uncommon.
    discoidea. F. Not uncommon.

MELIGETHES.
    rufipes. Gyll. Near Repton.
    æneus. F. Common.
    viridescens. F. Common.
    Kunzei. Er. Robin's Wood. Mr. Fowler.
    memnonius. Er. Repton.
    seniculus. Er. Foremark.
    picipes. Sturm. Very common.
    erythropus. Gyll. Robin's Wood.

POCADIUS.
    ferrugineus. F. Robin's Wood. Rare.

CYCHRAMUS.
    luteus. F. Bretby Wood.
    fungicola. Heer. Bretby Wood.

BYTURUS.
    sambuci. Scop. Common.
    tomentosus. F. Common.

CRYPTARCHA.
    strigata. F. Bretby Wood. Mr. Fowler.

IPS.
    quadripunctata. Hbst. Robin's Wood.

RHIZOPHAGUS.
    depressus. F. Robin's Wood.
    cribratus. Gyll. Robin's Wood.
    ferrugineus. Pk. Robin's Wood.
    dispar. Pk. Robin's Wood.
    bipustulatus. F. Robin's Wood.

COLYDIIDÆ.

CERYLON.
    histeroides. F. Under bark.

CUCUJIDÆ.

NAUSIBIUS.
    dentatus. Marsh. Scarce.

CRYPTOPHAGIDÆ.

TELMATOPHILUS.
    caricis. Ol. Near Repton.

ANTHEROPHAGUS.

nigricornis. F. Bretby Wood.

CRYPTOPHAGUS.

lycoperdi. Hbst. Puff balls. Bretby Wood.
saginatus. Sturm. Repton.
scanicus. L. Repton.
affinis. Sturm. Robin's Wood.
cellaris. Scop. Repton.
acutangulus. Gyll. Repton.
dentatus. Hbst. Robin's Wood.
vini. Pz. Common on furze.

PAREMECOSOMA.

melanocephala. Hbst. Scarce. In flood refuse.

ATOMARIA.

fimetarii. Hbst. Repton. Scarce.
nana. Er. Scarce.
umbrina. Gyll. Near Repton. Scarce.
pusilla. Pk. Haystack refuse. Common.
atricapilla. Steph. Common.
berolinensis. Kr. Not common.
fuscata. Schön. Not common.
mesomelas. Hbst. Common in osier-beds.
basalis. Er. Common in osier-beds.
nigripennis. Pk. Near Burton—Mr. Harris.
apicalis. Er. Scarce.
analis. Er. Scarce.
ruficornis. Marsh. Common.
versicolor. Er. In hotbeds.

EPHISTEMUS.
    globosus. Waltl. In hotbeds.
    gyrinoides. Marsh. In hotbeds.

MONOTOMA.
    spinicollis. Aub. In hotbeds.
    picipes. Pk. In hotbeds. Very common.
    brevicollis. Aub. In hotbeds. Rare.
    quadricollis. Aub. In hotbeds.
    longicollis. Gyll. In hotbeds.

LATHRIDIUS.
    lardarius. De G. Repton.
    angusticollis. Hum. Repton.
    ruficollis. Marsh. In haystack refuse.
    transversus. Ol. Repton.
    minutus. L. Repton.
    filum. Aub. Mr. Mason's Herbarium, Burton.
    nodifer. West. Repton.

CORTICARIA.
    punctulata. Marsh. Flood refuse.
    crenulata. Gyll. Flood refuse.
    denticulata. Gyll. Flood refuse.
    serrata. Pk. Flood refuse.
    elongata. Gyll. Common.
    gibbosa. Pk. Common.
    fuscula. Gyll. Common.

MYCETOPHAGIDÆ.

MYCETOPHAGUS.
    quadripustulatus. L. Scarce about Repton.
    multipunctatus. Hell. Scarce about Repton.

TRIPHYLLUS.
    punctatus. F. In fungi.
    suturalis. F. In fungi.

TYPHÆA.
    fumata. L. Haystack refuse.

MYCETÆA.
    hirta. Marsh. In corks of wine bottles.

DERMESTIDÆ.

DERMESTES.
    murinus. L. Dead animals in woods.
    lardarius. L. In rusted bacon.

ATTAGENUS.
    pellio. L. In houses. Repton.

ANTHRENUS.
    musæorum. L. Common.

BYRRHIDÆ.

BYRRHUS.
    pilula. L. Common.

CYTILUS.
    varius. F. Not uncommon.

SIMPLOCARIA.
    semistriata. Ill. Common.

PARNIDÆ.

MACRONYCHUS.
    4 tuberculatus. Müll. First discovered by Mr. Harris, in the Dove.

ELMIS.
    Volkmari. Pz. Milton Brook.
    parallelopipedus. Müll. Once in Milton Brook.

LIMNIUS.
    tuberculatus. Müll. Stenson brook. Common.

POTAMINUS.
    substriatus. Müll. In the Dove.

PARNUS.
    prolifericornis. F. By the Old Trent.
    auriculatus. Heer. By the Old Trent.

HETEROCERIDÆ.

HETEROCERUS.
    fusculus. Kies. By the Old Trent.

LUCANIDÆ.

LUCANUS.
    cervus. L. Once at Calke.

DORCUS.
    parallelopipedus. L. In old willows.

SINODENDRON.
    cylindricum. L. In old trees.

SCARABÆIDÆ.

ONTHOPHAGUS.
    cœnobita. Hbst. Occasionally.
    ovatus. L. Repton Waste.

APHODIUS.
erraticus. L. Not uncommon.
subterraneus. L. Common.
fossor. L. Common.
hæmorrhoidalis. L. Common.
scybalarius. F. Not uncommon.
fœtens. F. Scarce.
fimetarius. L. Very common.
ater. D. G. Very common.
granarius. L. Scarce.
inquinatus. F. Repton Waste.
pusillus. Hbst. Not uncommon.
merdarius. F. Common.
prodromus. Brahm. Common.
punctato-sulcatus. S. Common.
contaminatus. Hbst. Scarce.
rufipes. L. Common.
luridus. F. Common.
depressus. Kug. Not uncommon.

OXYOMUS.
porcatus. F. In hotbeds.

PSAMMODIUS.
sulcicollis. Ill. In hotbeds.

GEOTRUPES.
stercorarius. L. Common.
putridarius. Er. Common.

SERICA.
brunnea. L. Very rare about Repton.

MELOLONTHA.
    vulgaris. F. Common.

PHYLLOPERTHA.
    horticola. L. Occasionally common.

BUPRESTIDÆ.

AGRILUS.
    angustulus. Ill. Robin's Wood.
    laticornis. Ill. Robin's Wood.

EUCHNEMIDÆ.

MELASIS.
    buprestoides. L. Bretby Wood.

ELATERIDÆ.

LACON.
    murinus. L. Scarce about Repton.

CRYPTOHYPNUS.
    riparius. F. Common. By sweeping.
    quadripustulatus. F. By sweeping.
    dermestoides. Hbst. By sweeping.

MELANOTUS.
    rufipes. Hbst. Common.

LIMONIUS.
    minutus. L. Scarce about Repton.

ATHOUS.
    niger. L. Common.

hæmorrhoidalis. F. Common.
vittatus. F. Bretby Wood.
longicollis. Ol. Bretby Wood.

CORYMBITES.
cupreus. F. Occasionally about Repton.
tessellatus. L. Once, in flood refuse.
quercus. Gyll. Common.
v. ochropterus. Steph. Scarce.
holosericeus. F. Rare about Repton.

AGRIOTES.
sputator. L. Common.
lineatus. L. Common.
obscurus. L. Common.
sobrinus. Kies. Common.
pallidulus. Ill. Common.

DOLOPIUS.
marginatus. L. Bretby Wood.

CAMPYLUS.
linearis. L. Bretby Wood.

DASCILLIDÆ.

HELODES.
minuta. L. Common.
marginata. F. Near Burton. Mr. Harris.
livida. F. Common.

CYPHON.
coarctatus. Pk. Near Repton.
variabilis. Thun. Near Repton.
pallidulus. Boh. Near Repton.
padi. L. Near Repton.

PRIONOCYPHON.
    serricornis. Müll. Bretby Park. Mr. Harris.

SCIRTES.
    hemisphæricus. L. Rare near Repton.

TELEPHORIDÆ.

TELEPHORUS.
    alpinus. Pk. Bretby Wood.
    rusticus. Fall. Common.
    lividus. L. Not uncommon.
    lituratus. Fall. Common.
    figuratus. Mann. In Bull's meadow.
    pellucidus. F. Common.
    nigricans. Müll. Common.
    bicolor. F. Very common.
    thoracicus. Gyll. Osier-beds. Mr. Fowler.
    flavilabris. Fall. Very common.
    translucidus. Kry. Rare.
    fuscicornis. Ol. Not common.
    fulvus. Scop. Very common.
    testaceus. L. Rare.
    limbatus. Th. Common.
    pallidus. F. Bretby Wood.

MALTHINUS.
    fasciatus. Fall. Near Repton.
    balteatus. Suf. Near Repton.
    frontalis. Marsh. Near Repton.
    punctatus. Fourc. Near Repton.

MALTHODES.
>marginatus. Lat. Near Repton.
>fibulatus. Kies. Near Repton.
>dispar. Germ. Near Repton.
>flavoguttatus. Kies. Near Repton.
>sanguinolentus. Fall. Near Repton.
>misellus. Kies. Near Repton.

MALACHIUS.
>æneus. L. Rare about Repton.
>bipustulatus. L. Common.

ANTHOCOMUS.
>fasciatus. L. Occasionally about Repton.

DASYTES.
>plumbeus. Müll. In flowers, in gardens.
>plumbeo-niger. Goez. In flowers, in gardens.

CLERIDÆ.

TILLUS.
>elongatus. L. Once in my own house.

CORYNETES.
>cæruleus. De G. Common.
>ruficollis. F. Not common about Repton.
>violaceus. L. Scarce.

PTINIDÆ.

HEDOBIA.
>imperialis. L. Scarce about Repton.

PTINUS.
>sexpunctatus. Pz. Houses. Repton and Newton Solney.
>subpilosus. Müll. Houses. Repton.
>fur. L. Houses. Repton.

NIPTUS.
>hololeucus. Fald. Houses. Repton.
>crenatus. F. Houses. Repton.

PRIOBIUM.
>castaneum. F. Repton.

ANOBIUM.
>domesticum. Four. Houses. Repton.
>paniceum. L. Houses. Repton.

XESTOBIUM.
>tessellatum. F. Repton.

PTILINUS.
>pectincornis. L. Old willows about Repton.

OCHINA.
>hederæ. Müll. The Hayes, Repton.

BOSTRYCHIDÆ.

LYCTUS.
>canaliculatus. F. On palings.

CISSIDÆ.

CIS.
>boleti. Scop. In boleti.
>villosulus. Marsh. In boleti.

micans. Hbst. In boleti.
hispidus. Pk. In boleti.
bidentatus. Ol. In boleti.
vestitus. Mel. In boleti.

OCTOTEMNUS.
glabriculus. Gyll. In boleti.

TENEBRIONIDÆ.

BLAPS.
mortisaga. L. In Mr. Worthington's Brewery.
mucronata. Latr. Cellars and outhouses.

SCAPHIDEMA.
æneum. Pk. Old willows, and flood refuse.

TRIBOLIUM.
ferrugineum. F. Corn granaries. Mr. Harris.
confusum. Duv. Corn granaries. Mr. Harris. Rare.

TENEBRIO.
molitor. L. In bakehouses.

PYTHIDÆ.

SALPINGUS.
castaneus. Pz. By sweeping.
foveolatus. Ljun. Robin's Wood. Flood refuse.

LISSODEMA.
4 pustulata. Marsh. By sweeping.

RHINOSIMUS.
ruficollis. L. Under bark.
viridipennis. Steph. Under bark.
planirostris. F. Under bark.

MELANDRYADÆ.

TETRATOMA.
    fungorum. F. Near Ridgeway.

ORCHESIA.
    micans. Pz. In boleti.
    minor. Walk. Repton Shrubs. Very rare.

MELANDRYA.
    caraboides. L. In dead trees and posts.

CONOPALPUS.
    testaceus. Ol. Bretby Park and Wood.

ANTHICIDÆ.

ANTHICUS.
    floralis. L. Hotbeds. Very common.
    antherinus. L. By sweeping.

PYROCHROIDÆ.

PYROCHROA.
    serraticornis. Scop. Repton.

MORDELLIDÆ.

ANASPIS.
    frontalis. L. In flowers about Repton.
    rufilabris. Gyll. In Bretby Wood.
    forcipata. Muls. Rare.
    fasciata. Forst. Common in flowers.
    ruficollis. F. Common in flowers.
    thoracica. L. In flowers.
    subtestacea. Steph. In flowers.
    melanopa. Forst. In flowers.

RHIPIDOPHORIDÆ.

METŒCUS.
    paradoxus. L. In wasps' nests.

CANTHARIDÆ.

MELOE.
    proscarabæus. L. Repton.
    violaceus. Marsh. Robin's Cross.

ŒDEMERIDÆ.

ASCLERA.
    cœrulea. L. Repton.

CURCULIONIDÆ.

OTIORHYNCHUS.
    ligneus. Ol. Near Repton.
    picipes. F. Common.
    sulcatus. F. Near Repton.
    ovatus. L. Near Repton.
    muscorum. Bris. By the Dove.

OMIAS.
    mollinus. Boh. Bull's in the Meadow.

BARYPEITHES.
    brunnipes. Ol. Common.

PLATYTARSUS.
    echinatus. Bons. By sweeping.

PHYLLOBIUS.
    calcaratus. F. Local—but common on alders.
    alneti. F. Very common on nettles.
    pyri. L. Bretby Wood.
    argentatus. L. Common.
    maculicornis. Germ. Bretby Wood—not common.
    oblongus. L. Common.
    pomonæ. Ol. Common.
    uniformis. Marsh. Common.

TROPIPHORUS.
    mercurialis. F. Repton Shrubs.

CNEORHINUS.
    exaratus. Marsh. Lane near Burton. Mr. Harris.

LIOPHLŒUS.
    nubilus. F. Not uncommon.

BARYNOTUS.
    obscurus. F. Repton Shrubs.
    mœrens. F. Near Repton.

STROPHOSOMUS.
    coryli. F. Common.
    obesus. Marsh. Bretby Wood.
    retusus. Marsh. Scarce about Repton.
    faber. Hbst. Findern.

SITONES.
    flavescens. Marsh. Near Repton.
    suturalis. Steph. Bull's, in the Meadow.
    sulcifrons. Thun. Common.
    tibialis. Hbst. Bull's, in the Meadow.

cambricus. Steph. Bretby Wood. Mr. Fowler.
regensteinensis. Hbst. Stenson gorse.
puncticollis. Steph. Fields about Repton.
lineatus. L. Very common.
hispidulus. F. Common.
humeralis. Steph. Bull's, in the Meadow.

POLYDROSUS.
undatus. F. Common.
pterygomalis. Soh. Common in woods.
cervinus. L. Bretby Wood.
micans. F. Bretby Wood.

SCIAPHILUS.
muricatus. Fab. Bretby wood.

LIOSOMUS.
ovatulus. Clair. Bretby Wood.

ALOPHUS.
triguttatus. F. Not uncommon.

HYPERA.
punctata. F. Common.
rumicis. L. Near Repton.
suspiciosa. Hbst. Merrybower.
plantaginis. De. G. Near Repton.
variabilis. Hbst. Near Repton.
nigrirostris. F. Common.

HYLOBIUS.
abietis. L. Scarce about Repton.

GRYPIDIUS.
equiseti. F. Egginton.

ERIRHINUS.
bimaculatus. F. Flood refuse.
acridulus. L. Common.
nereis. Pk. Willington Pits.
vorax. F. In poplars.
maculatus. Marsh. On sallows, Bretby Wood.
tortrix. L. On aspens, Bretby Wood.

MECINUS.
pyraster. Hbst. Not uncommon.

HYDRONOMUS.
alismatis. Marsh. On watercress.

BAGOUS.
diglyptus. Boh. New to Great Britain. Discovered by Mr. Harris, near Burton.

TANYSPHIRUS.
lemnæ. F. On duckweed.

ANOPLUS.
plantaris. Nætz. Bretby Wood.

BALANINUS.
glandium. Marsh. Bretby Wood.
nucum. L. Bretby Wood.
tessellatus. Four. Bretby Wood.
villosus. Hbst. Bretby Wood.
brassicæ. Fab. Common.
pyrrhoceras. Marsh. Common.

ANTHONOMUS.
ulmi. De. G. Near Repton.
pedicularius. L. Near Repton.

pomorum. L. Near Repton.
rubi. Hbst. Common.

ORCHESTES.
quercus. L. Common.
ferrugineus. Marsh. Near Repton.
alni. L. Near Repton.
ilicis. F. Near Repton.
fagi. L. Near Repton.
rusci. Hbst. Bretby Wood.
stigma. Germ. On sallows.
saliceti. F. On sallows.
salicis. L. On sallows.

RAMPHUS.
flavicornis. Clair. Bretby Wood.

ELLESCHUS.
bipunctatus. L. Bretby Wood. Rare.

TYCHIUS.
meliloti. Steph. Rare near Repton.
picirostris. F. Common.

CIONUS.
scrophulariæ. L. Bretby Wood.
verbasci. F. Bretby Wood.
hortulanus. Marsh. Bretby Wood.
blattariæ. F. Bretby Wood.
pulchellus. Hbst. Bretby Wood.

NANOPHYES.
lythri. F. In osier-beds.

GYMNETRON.
pascuorum. Gyll. Fields near Repton.
villosulus. Gyll. On water nasturtium.
beccabungæ. L. On water nasturtium.
noctis. Hbst. On toad-flax, Twyford Road.

OROBITIS.
cyaneus. L. Scarce, at Repton Shrubs.

ACALLES.
roboris. Curt. Robin's Wood.
ptinoides. Marsh. Robin's Wood.
turbatus. Boh. Robin's Wood.

CRYPTORHYNCHUS.
lapathi. L. Osier-beds.

CŒLIODES.
quercus. F. Common on oak.
ruber. Marsh. Robin's Wood.
rubicundus. Pk. Robin's Wood.
subrufus. Hbst. Robin's Wood.
quadrimaculatus. L. Very common on nettles.
fuliginosus. Marsh. Not uncommon.

CEUTHORYHNCHUS.
assimilis. Pk. Common.
erysimi. F. Common.
contractus. Marsh. Common.
cochleariæ. Gyll. Fields near Repton.
viduatus. Gyll. Rare. Robin's Wood.
litura. F. Robin's Wood.
rugulosus. Hbst. Twyford.
melanostictus. Marsh. Near Repton.

quadridens. Pz. Common.
pollinarius. Forst. Common on nettles.
sulcicollis. Gyll. Common.
cyanipennis. Germ. Findern.
chalybæus. Germ. Findern.

CEUTHORHYNCHIDEUS.
floralis. Pk. Near Repton.
hepaticus. Gyll. Near Repton.
nigrinus. Marsh. Near Repton.
pyrrorhynchus. Marsh. Near Repton.
melanarius. Steph. Near Repton.
terminatus. Hbst. Near Repton.
troglodytes. F. Near Repton. Common.
versicolor. Bris. Near Repton.

AMALUS.
scortillum. Hbst. Robin's Wood.

POOPHAGUS.
sisymbrii. F. On watercress.

PHYTOBIUS.
velatus. Beck. By the Old Trent.
leucogaster. Marsh. By the Old Trent.

RHINONCHUS.
pericarpius. F. Common.
subfasciatus. Gyll. Fields near Repton.
Castor. F. Fields near Repton.

BARIS.
T-album. L. Not uncommon.

CALANDRA.
    granaria. L. In granaries.

MAGDALINUS.
    cerasi. L. Bretby Wood.
    pruni. L. Bretby Wood.

APION.
    pomonæ. F. Common.
    carduorum. Kirb. Common.
    onopordi. Kirb. Common.
    ulicis. Forst. Stenson Gorse.
    pallipes. Kirb. Robin's Wood.
    æneum. F. Common.
    radiolus. Kirb. Common.
    striatum. Kirb. Stenson Gorse.
    simile. Kirb. Rare. Bretby Wood.
    seniculum. Kirb. By sweeping.
    rufirostre. F. Common.
    viciæ. Pk. Common.
    varipes. Germ. Scarce about Repton.
    fagi. L. Common.
    assimile. Kirb. Common.
    trifolii. L. Common.
    flavipes. F. Common.
    nigritarse. Kirb. Common.
    ebeninum. Kirb. Scarce about Repton.
    punctigerum. Pk. Scarce about Repton.
    virens. Hbst. Common.
    platalea. Germ. Not uncommon.
    ervi. Kirb. Common.
    ononis. Kirb. Common.

minimum. Hbst. Bretby Wood.
pisi. F. Common.
æthiops. Hbst. Near Repton.
scutellare. Kirb. Stenson Gorse.
meliloti. Kirb. Rare about Repton.
loti. Kirb. Rare about Repton.
Spencei. Kirb. Not uncommon.
vorax. Hbst. Not uncommon.
miniatum. Germ. On docks. Rare about Repton.
malvæ. F. Rare about Repton.
violaceum. Kirb. Not uncommon.
hydrolapathi. Kirb. Not uncommon.
humile. Germ. Very common.

RHYNCHITES.
betuleti. F. Bretby Wood.
æquatus. L. Near Repton.
æneovirens. Marsh. Near Repton.
conicus. Ill. Near Repton.
alliariæ. Pk. Near Repton.
germanicus. Hbst. Near Repton.
nanus. Pk. Bretby Wood.
pubescens. Hbst. Bretby Wood.
megacephalus. Germ. Bretby Wood.
betulæ. L. Bretby Wood.

ATTELABUS.
curculionoides. L. Robin's Wood. Rare.

APODERUS.
coryli. L. Bretby Wood. Rare.

SCOLYTIDÆ.

HYLASTES.
    ater. Pk. Near Repton.
    opacus. Er. Near Repton.

HYLURGUS.
    piniperda. L. Repton Rocks.

HYLESINUS.
    crenatus. F. Near Repton.
    oleiperda. F. Near Repton.
    fraxini. F. Near Repton.
    vittatus. F. Near Repton.

SCOLYTUS.
    destructor. Ol. Near Repton.

CRYPHALUS.
    abietis. Ratz. Bretby Wood.

DRYOCŒTES.
    villosus. F. Bretby Wood.

PITYOPHTHORUS.
    micrographus. Gll. Near Repton.
    bidens. F. Robin's Wood.

XYLOCLEPTES.
    bispinus. Duft. Robin's Wood.

TOMICUS.
    acuminatus. Gyll. Robin's Wood.
    laricis. F. Robin's Wood.

ANTHRIBIDÆ.

BRACHYTARSUS.
    scabrosus. F. Bretby Wood.
    varius. F. Bretby Wood.

CHORAGUS.
    Sheppardi. Kirb. Near Newton Road.

BRUCHIDÆ.

BRUCHUS.
    rufimanus. Boh. In pea fields.

CERAMBYCIDÆ.

AROMIA.
    moschata. L. Osier-beds.

CALLIDIUM.
    violaceum. L. Near Burton.
    alni. L. Robin's Wood.
    variabile. L. Robin's Wood.

CLYTUS.
    arietis. L. Common on palings.
    mysticus. L. Repton.

GRACILIA.
    pygmæa. F. Repton.

MONOHAMMUS.
    sartor. F. Repton. Mr. Brown.
    sutor. L. Burton. Mr. Brown.

LIOPUS.
    nebulosus. L. Robin's Wood.

POGONOCHERUS.
    hispidus. L. Bretby Wood.
    dentatus. Four. Bretby Wood.

SAPERDA.
    scalaris. L. Bretby Wood. Once by Mr. Fowler.
    populnea. L. Robin's Wood.

POLYOPSIA.
    præusta. L. Near Repton.

STENOSTOLA.
    ferrea. Schr. Bretby Wood.

PHYTŒCIA.
    cylindrica. L. Once near Repton.

RHAGIUM.
    inquisitor. F. Near Repton.
    indagator. L. Near Repton.
    bifasciatum. F. Near Repton.

TOXOTUS.
    meridianus. L. Bretby Wood.

STRANGALIA.
    armata. Hbst. Bretby Wood.
    melanura. L. Bretby Wood. Mr. Brown.

GRAMMOPTERA.
    ruficornis. F. Near Repton.

CHRYSOMELIDÆ.

DONACIA.
    crassipes. F. Near Burton.
    sparganii. Ahr. Near Burton.

dentipes. F. Bull's, in the Meadow.
lemnæ. F. Bull's, in the Meadow.
impressa. Pk. Bretby Park and Lane.
linearis. Hoppe. Common.
semicuprea. Pz. Common.
menyanthidis. F. Near Burton.
sericea. L. Common.

HÆMONIA.
equiseti. F. Near Burton.

ZEUGOPHORA.
subspinosa. F. Bretby Wood.

LEMA.
cyanella. F. Common.
melanopa. L. Common.

CRYPTOCEPHALUS.
pusillus. F. Bretby Wood.
labiatus. L. Bretby Wood.

TIMARCHA.
lævigata. L. Common.
coriaria. F. Common.

CHRYSOMELA.
staphylæa. L. Common.
gœttingensis. L. Near Burton.
menthastri. Suf. Bretby Park.
fastuosa. L. Knoll Hills.
polita. L. Common.
hyperici. Forst. Robin's Wood.
didymata. Scrib. Robin's Wood.

LINA.
 ænea. F. Robin's Wood. Mr. Fowler.

GONIOCTENA.
 rufipes. Gyll. Bretby Wood.
 litura. F. Stenson.
 pallida. L. Bretby Wood. Mr. Brown.

GASTROPHYSA.
 polygoni. L. Common.
 raphani. F. Common.

PHÆDON.
 tumidulum. Kirb. Common.
 betulæ. L. Local.
 cochleariæ. F. Common on horse radish.

PHRATORA.
 vulgatissima. L. Common.
 cavifrons. Th. Bretby Wood.
 vitellinæ. L. Common.

PRASOCURIS.
 aucta. F. Near Repton.
 marginella. L. Near Repton.
 phellandrii. L. Near Repton.
 beccabungæ. Ill. Near Repton.

ADIMONIA.
 capreæ. L. Common.
 sanguinea. F. Ingleby. On whitethorn.

GALERUCA.
 lineola. F. Osier-beds.

sagittariæ. Gyll. Crewe's Pond.
viburni. Pk. Robin's Wood.

AGELASTICA.
halensis. L. Robin's Wood.

PHYLLOBROTICA.
quadrimaculata. L. Near Burton.

LUPERUS.
betulinus. Fourc. Near Repton.
flavipes. L. Near Repton.

HALTICA.
ericeti. All. Repton Rocks.
pusilla. Duft. Repton Rocks.

CREPIDODORA.
transversa. Marsh. Common.
ferruginea. Scop. Common.
rufipes. L. Scarce about Repton.
aurata. Marsh. Common.
chloris. Fond. Robin's Wood.
Modeeri. L. Bretby Wood.
salicariæ. Pk. Bretby Wood.
ventralis. Ill. Bretby Wood.

MANTURA.
rustica. L. Bretby.
obtusata. Gyll. Once, in flood refuse.

BATOPHILA.
rubi. Pk. Findern.
ærata. Marsh. Findern.

APHTHONA.
lutescens. Gyll. Bretby Wood.
nigriceps. Redt. Egginton.
cœrulea. Pk. Robin's Wood.

PHYLLOTRETA.
nodicornis. Marsh. Ticknall Quarry.
melæna. Pk. Repton. On gilliflowers.
obscurella. Ill. Repton. On gilliflowers.
punctulata. Marsh. Repton. On gilliflowers.
vittula. Redt. Not uncommon.
undulata. Kuts. Very common.
nemorum. L. Not uncommon.
sinuata. Steph. Bretby Wood.
brassicæ. F. Not uncommon.

PLECTROSCELIS.
concinna. Marsh. Common.
aridella. Pk. Not uncommon.

THYAMIS.
holsatica. L. Bretby Wood.
lurida. Scop. Common.
fuscicollis. Steph. Common.
atricilla. Gyll. Common.
melanocephalus. Gyll. Common.
suturalis. Marsh. Near Repton.
nasturtii. F. Near Burton.
Foudrasi. Crotch. Near Repton.
pusilla. Gyll. Near Repton.
ochroleuca. Marsh. Ticknall.
lævis. Duft. Near Repton.

pellucida. Foud. Near Repton.
flavicornis. Steph. Near Repton.
teucrii. All. Near Repton.

PSYLLIODES.
chrysocephala. F. Near Repton.
cupro-nitens. Forst. Foremark Bottoms.
affinis. Pk. Near Repton.
picina. Marsh. Robin's Wood.

APTEROPEDA.
graminis. Pz. Bretby Wood.
globosa. Pz. Bretby Wood.

SPHÆRODERMA.
testacea. F. Common.
cardui. Gyll. Common.

CASSIDA.
viridis. L. Common on thistles.
sanguinolenta. F. Loscoe and Repton Shrubs.
nobilis. L. Near Repton.
obsoleta. Ill. Near Repton.
equestris. F. Scarce about Repton.
hemisphærica. Hbst. Near Burton. Mr. Harris.

EROTYLIDÆ.

ENGIS.
humeralis. F. Bretby.
rufifrons. F. Bretby.

TRIPLAX.
russica. L. Scarce near Repton.

COCCINELLIDÆ.
COCCINELLA.
- 19 punctata. L. Willington Pits.
- obliterata. L. Repton Rocks.
- bipunctata. L. Common.
- 7 punctata. L. Common.
- variabilis. Ill. Common.
- 18 guttata. L. Repton Rocks.
- 14 guttata. L. Repton Rocks.
- 14 punctata. L. Common.
- 22 punctata. L. Common.

SCYMNUS.
- nigrinus. Kug. Near Repton.
- discoideus. Ill. Near Repton.
- Mulsanti. Wat. Near Repton.
- capitatus. F. Near Repton.

RHIZOBIUS.
- litura. F. Common.

COCCIDULA.
- rufa. Hbst. Common.
- scutellata. Hbst. Rare about Repton.

ALEXIA.
- pilifera. Müll. Bretby Wood.

A LIST OF LEPIDOPTERA

OBSERVED AT AND NEAR

REPTON AND WILLINGTON, DERBYSHIRE.

RHOPALOCERA.

Pieris Brassicæ.
,, *Rapæ.* Yellowish varieties occasionally in September.
,, *Napi.* Very small variety of the female at Milton.
Anthocaris Cardamines. One dark variety of the male.
Gonepteryx Rhamni. Principally the autumnal broods. Clover fields.
Colias Edusa. Clover fields, Repton. One variety Helice.
Chrysophanus Phlœas.
Polyommatus Argiolus. Scarce. Near hollies, in April.
,, *Alexis.*
Argynnis Paphia. Repton Shrubs and Anchor Church; but not in late years.
Argynnis Euphrosyne. Formerly abundant in Repton Shrubs.
,, *Selene.* Ditto.
Vanessa Cardui. Occasionally in Autumn, or hybernated.
,, *Atalanta.*

Vanessa Io.
,, *Urticæ.* Only this year (1865) plentiful, since the hard winter of 1860-61.
Vanessa Polychloros. Scarce. Near Hartshorne and Calke.
,, *C. Album.* On ripe fruit and Michaelmas daisies.
,, *Antiopa.* One near Milton.
Satyrus Megæra. Scarce since 1861.
,, *Ægina.* Scarce. Repton Shrubs, flying in the shade.
,, *Janira.*
,, *Tithonus.*
,, *Hyperanthus.* Repton Shrubs, and Findern Covert, but not lately.
Satyrus Pamphilus. One in a meadow near Repton.
Thecla W. Album. Repton. (P. B. M.)
Pamphila Sylvanus. Occasionally at The Oaks, near Burton.

HETEROCERA.

SPHINGINA.

Trochilium Cynipiforme. Repton Shrubs, on stumps of trees (chiefly females).
Trochilium Tipuliforme. On Currant trees.
,, *Sphegiforme.* One or two specimens in Alder, Repton Shrubs.
Ægeria Apiformis. Osier beds at Repton.
,, *Bembiciformis.* Larvæ in Poplars in Findern Covert.
Macroglossa Stellatarum.
Chærocampa Elpenor. Scarce. Larvæ in wet places.
Sphinx Convolvuli. Repton, some years ago in considerable abundance.

Sphinx Ligustri.
Acherontia Atropos. Occasionally in some abundance.
Smerinthus Ocellatus.
,, *Populi.*
Anthrocera Lonicerœ. In mowing grass, on Burnet, &c.
,, *Filipendulæ.* Ditto.
Procris Statices. In mowing-grass near Milton.

BOMBYCINA.

Lithosia Complanula.
Nudaria Mundana. On nut.
Arctia Caja.
Phragmatobia Fuliginosa. Once at Willington.
,, *Lubricipeda.*
,, *Menthrasti.*
Liparis Auriflua.
,, *Salicis.* On Poplars at Findern.
Orgyia Pudibunda. Scarce.
,, *Antiqua.*
Pœcilocampa Populi. Near Repton. (P. B. M.)
Eriogaster Lanestris.
Trichiura Cratægi. Larvæ on the highest shoots of Hawthorn.
Lasiocampa Quercus.
Odonestis Potatoria.
Cossus Ligniperda.
Hepialus Humuli.
,, *Velleda.* One in Repton Shrubs.
,, *Lupulinus.* Several varieties of the female.
,, *Hectus.* Repton Shrubs.
Cilix Spinula.

Cerura Furcula. Near Repton. (P. B. M.)
„ *Bifida.*
„ *Vinula.*
Notodonta Camelina. Rare.
„ *Dictæa.* Near Repton. (P. B. M.)
„ *Ziczac.* Near Repton. (P. B. M.)
„ *Chaonia.* Near Repton. (P. B. M.)
„ *Dodonea.* Near Repton. (P. B. M.)
Diloba Cœruleocephala.
Pygæra Bucephala.

NOCTUINA.

Semaphora Tridens.
Apatela Leporina. One in the Potlock Covert.
Acronycta Alni. Two Larvæ.
„ *Megacephala.*
Ceropacha Diluta. Gorstey Leys, near Ingleby.
Cymatophora Viminalis.
Bryophila Perla.
Stilbia Anomala. One, Findern.
Caradrina Morpheus.
„ *Cubicularis.*
Grammesia Trilinea and variety *Bilinea.*
Leucania Lithargyria.
„ *Comma.*
„ *Impura.*
„ *Pallens.*
Nonagria Despecta ?
„ *Fulva.*
„ *Typhæ.* Abundant in the larvæ state, feeding inside Typha.

Gortyna Flavago.
Hydrecia Micacea.
Axylia Putris. Near Repton. (P. B. M.)
Miana Literosa. Willington.
„ *Fasciuncula.* All the varieties.
„ *Strigilis.* Ditto.
„ *Arcuosa.*
Apamea Didyma. All the varieties.
„ *Unanimis ?*
„ *Gemina* and variety *Oblonga.*
Luperina Testacea.
„ *Basilinea.*
„ *Infesta.*
Xylophasia Scolopacina. One at Gorstey Leys, near Ingleby.
„ *Hepatica.*
„ *Rurea.* In varieties.
„ *Lythoxylea.*
„ *Sublustris.* One in the Potlock Covert.
„ *Polyodon.* Dark and light varieties.
Triphœna Pronuba.
„ *Orbona.*
„ *Fimbria.* One in the Potlock Covert at Sugar. One on the Etwall road in September, which laid eggs, hatched about three weeks afterwards.
Triphœna Janthina. Flying frequently by day.
„ *Interjecta.* Ditto. Scarce.
Segetia Xanthographa.
Rusina Tenebrosa. One in Repton Shrubs at Sugar.
Noctua Umbrosa. Larvæ on seeds of Wild Hyacinth.
„ *Bella.*
„ *Baja.*

Noctua Festiva. Potlock Covert at Sugar.
,, *Brunnea.* Ditto.
,, *Triangulum.*
,, *C. Nigrum.*
Chersotis Plecta.
Spœlotis Augur.
Agrotis Saucia. One in the Potlock Covert at Sugar.
,, *Suffusa.* The Henhurst, near Burton.
,, *Segetum.* In varieties.
,, *Exclamationis.* In varieties.
,, *Aquelina.* One.
,, *Putris.*
Cerapteryx Graminis.
Heliophobus Popularis. Female scarce, at rest on Grass.
Tœniocampa Gothica.
,, *Stabilis.*
,, *Instabilis.*
,, *Gracilis.* One on Willow blossom at Willington.
,, *Munda.* Near Repton. (P. B. M.)
,, *Cruda.*
Orthosia Upsilon. At Sugar on Willows, etc., near the Trent.
,, *Lota.*
Anthrocelis Litura. Findern Covert.
,, *Pistacina.* Ditto.
Scoliopteryx Libatrix.
Cosmia Diffinis. Etwall.
Euperia Trapezina. Larvæ dangerous in Breeding cages, as eating other larvæ.
Cirredia Xerampelina. One at rest on Willow, near Willington Brook. (Not rare at Barrow-on-Trent. W. G.)
Xanthia Ferruginea.

Xanthia Rufina. The Henhurst, near Burton-on-Trent.
,, *Gilvago.*
,, *Silago.*
,, *Cerago.*
Glœa Spadicea. On Ivy blossoms at Willington.
,, *Vaccinii.* Ditto.
Scopelosoma Satellitia. Larvæ have the same habit as those of *Euperia Trapezina.*
Xanthia Citrago. Near Repton. (P. B. M.)
Miselia Oxyacanthæ.
Chariptera Aprilina.
Dianthœcia Capsincola.
,, *Cucubali.*
Polia Chi.
Hadena Brassicæ.
,, *Adusta.* Potlock Covert.
,, *Suasa.*
,, *Oleracea.*
,, *Dentina.*
,, *Protea.*
,, *Chenopodii.*
Aplecta Nebulosa. At rest on trunks of trees in Repton Shrubs.
Aplecta Herbida. The Potlock Covert.
Phlogophora Meticulosa.
Euplexia Lucipara.
Thyatira Batis. Potlock Covert, at Sugar.
,, *Derasa.*
Calocampa Exoleta. Potlock Covert, and hybernated.
Cucullia Chamomillœ. One bred from Larva found at Willington.

Cucullia Umbratica.
Xylocampa Lithoriza. On Sallow blossoms.
Heliodes Heliaca. On daisies in the day-time.
Plusia Gamma.
,, *Iota.*
,, *Inscripta.*
,, *Festucæ.* Near Derby.
,, *Chrysitis.*
Abrostola Urticæ.
,, *Triplasia.*
Nœnia Typica. Larvæ on Larkspur, etc.
Mania Maura.
Amphipyra Pyramidea. In my own garden at Sugar (W. G.)
,, *Tragopogonis.*
Catocala Fraxini. Burton. (P. B. M.)

PYRALIDINA.

Pyrausta Cespitalis. Repton Park.
Pyralis Farinalis.
,, *Glaucinalis.* In abundance at Willington, 1865. Larvæ feeding on dried leaves, cocoons very remarkable.
Aglossa Pinguinalis.
Hydrocampa Lemnalis.
,, *Stratiotalis.* At the "Water-meetings," Repton Meadows.
Hydrocampa Nymphæalis.
,, *Potomogalis.*
Acentropus Niveus. On the Trent at Willington.
Ebulia Sambucalis.
Scopula Prunalis.

Scopula Olivalis.
,, *Etialis.*
,, *Ferrugalis.* One at Willington, near an Osier-bed. October, 1865.
Botys Fuscalis. Scarce. In mowing grass, in damp places.
,, *Urticalis.*
,, *Verticalis.* At Willington, but scarce. Abundant at "The Oaks," near Burton-on-Trent.
Stenopteryx Hybridalis. In pasture Fields, late in the summer.
Polypogon Barbalis.
,, *Grisealis.*
Hypena Proboscidalis.
Nola Cucullalis. Larvæ small, bristly, grey, feeding on Hawthorn. Pupæ spin up in curious triangular cases on the branches.
Nola Cristulalis. At rest on trunks of trees in Repton Shrubs. Scarce.
Schœnobius Forficellus. Railway Cuttings at Willington.
,, *Gigantellus.* Near Repton. (P. B. M.)
Crambus Perlellus. Near Repton. (P. B. M.)
,, *Pratellus.*
,, *Pascuellus.*
,, *Hortellus.*
,, *Culmellus.*
,, *Tristellus.*
Eudorea Ambigualis.
,, *Pyralella.*
,, *Frequentella.*
,, *Pallida.* In Railway Cuttings.
Aphomia Colonella.

Achroea Grisella. About beehives at Tickenall.
Ephestia Elutella. Amongst bundles of dried herbs. Druggists' shops, Derby.
Ephestia Interpunctella. In Grocers' warehouses, Derby.
Acrobasis Consociella. One near Newhall.
Cryptoblabes Bistriga. Repton Shrubs, on trunks of trees.
Myelois Artemisiella. At Etwall Hall.

PTEROPHORINA.

Pterophorus Fuscodactylus.
,, *Bipunctidactylus.*
,, *Pentadactylus.*
,, *Ochrodactylus.*
,, *Lithodactylus.* Near Burton-on-Trent.
,, *Acanthodactylus.* One at Willington.
Alucita Polydactyla.

GEOMETRINA.

Geometra Papillionaria. Repton and near Willington Brook. Scarce.
Hemithea Cythisaria. One at Willington.
Chlorochroma Æruginaria.
,, *Æstivaria.*
Metrocampa Margaritaria. Female scarce.
Ourapteryx Sambucaria.
Rumia Crategaria Pupæ light green.
Pericallia Syringaria. Frequenting Nut trees. Scarce.
Epione Apicaria. Osier beds at Repton.
Eurymene Dolobraria. Near Repton. (P. B. M.)
Ennomos Illunaria. On Sallow blossom.
,, *Lunaria.* One at Repton Shrubs.

Ennomos Angularia.
,, *Tiliaria.*
,, *Fuscantaria.* One near Egginton.
Odontopera Bidentaria. Repton Shrubs.
Crocallis Elinguaria. Larvæ on plum trees.
Himera Pennaria.
Halia Wavaria. Larvæ abundant on Red Currants.
Asthena Sylvata. Once, near Repton. (P. B. M.)
,, *Blomeraria.* Once, near Repton. (P. B. M.)
Eupisteria Hepararia. Repton Shrubs.
Anisopteryx Æscularia.
Hibernia Leucophœaria. Repton Shrubs, on trunks of trees in March.
Hibernia Rupicapraria.
,, *Progemmaria.*
,, *Aurantiaria.* One larva in Repton Shrubs.
,, *Defoliaria.*
Phigalia Pilosaria. Rolleston. Pupæ at roots of trees.
Biston Prodromaria. One, Findern Covert.
,, *Betularia.* Repton Shrubs, etc.
Boarmia Repandaria. ditto.
,, *Rhomboidaria.* On Walls, etc., at rest.
Hemerophila Abruptaria. Abundant at Willington about ten years ago in May. I have not seen a specimen here since that time.
Cleora Lichenaria.
Tephrosia Crepuscularia. At rest on trees in Repton Shrubs.
,, *Punctulata.* Near Repton. (P. B. M.)
Fidonia Piniaria. One in the Potlock Covert, Willington.
Lozogramma Petraria. Frequenting Fern (Pteris), Repton Shrubs and Potlock Covert.

Eubolia Cervinaria. Best found in the larvæ state, on Mallow.
Eubolia Mensuraria.
Coremia Didymaria.
„ *Ferrugaria.*
„ *Unidentaria.* Sometimes apparently flying gregariously.
Coremia Pectinitaria.
„ *Montanaria.*
„ *Fluctuaria.*
„ *Propugnaria.*
Thera Simularia. On Spruce Firs in the Potlock Covert.
Anticlea Derivaria. Generally at rest on marshy banks; on Sallow blossom.
Anticlea Badiaria. On Sallows.
Steganolophia Ribesiaria. In Gardens. August.
Harpalyce Suffumaria. Repton Rocks.
„ *Silacearia.*
„ *Ocellaria.*
„ *Fulvaria.*
„ *Marmoraria.*
„ *Pyraliaria.*
„ *Russaria.*
„ *Immanaria.*
Ypsipetes Elutaria. Larvæ reddish, in catkins of Sallow when in seed.
Ypsipetes Impluviaria.
Phœsyle Miaria. Scarce.
Chematobia Dilutaria.
„ *Brumaria.* Female best obtained by breeding from the larvæ.

Triphosa Dubitaria.
„ *Certaria ?*
Phibalapteryx Lignaria. For two successive years in some wet ground by the side of the Railway at Willington, but not of late years.
Camptogramma Bilinearia.
Cidaria Corylata. Near Repton. (P. B. M.)
„ *Fulvata.* Near Repton. (P. B. M.)
„ *Pyraliata.* Near Repton. (P. B. M.)
„ *Dotata.* Near Repton. (P. B. M.)
Pelurga Comitata. Near Repton. (P. B. M.)
Venilia Macularia. Many years ago one wing found at Willington. (Common in Dovedale.)
Melanippe Alchemillaria.
Emmelesia Rivularia.
„ *Hydraria.*
„ *Decoloraria.*
„ *Albularia.*
Zerene Albicillaria.
„ *Rubiginaria.* On Alders, Bretby Brook.
Abraxas Grossularia.
„ *Ulmaria.* In woods under Elms. Both this and the preceding species seem singularly free from varieties.
Bapta Temeraria. Scarce.
Cabera Pusaria.
„ *Exanthemaria.*
Ephyra Punctaria. Repton Shrubs, at rest on leaves.
Eupithecia Pulchellata. Near Repton. (P. B. M.)
„ *Valerianata.* Near Repton. (P. B. M.)
„ *Trisignata.* Near Repton. (P. B. M.)

Eupithecia Albipunctata. Near Repton. (P. B. M.)
,, *Fraxinata.* Near Repton. (P. B. M.)
,, *Subnotata.* Near Repton. (P. B. M.)
,, *Vulgata.* Near Repton. (P. B. M.)
,, *Assimilata.* Near Repton. (P. B. M.)
,, *Exiguata.* Near Repton. (P. B. M.)
,, *Rectangularia.*
,, *Minutaria.* Bred for larvæ on Wormwood.
,, *Plumbeolaria ?*
,, *Abbreviaria.*
,, *Innotaria.*
,, *Castigaria.*
,, *Austeraria.*
,, *Subfulvaria.* Larvæ in profusion on Yarrow in October, on Railway Banks.
Lobophora Hexapterata. Near Repton. (P. B. M.)
Dosithea Virgularia. Beaten out of Ivy.
,, *Scutularia.*
,, *Reversaria.*
Acidalia Luteraria. Repton Shrubs.
,, *Silvaria.* Repton Shrubs, but not lately.
,, *Remutaria.* Henhurst. Repton Shrubs.
,, *Candidaria.*
,, *Aversaria,* and in varieties.
Pœcilophasia Marginaria.
Timandra Imitaria.
Ania Emarginaria.
Bradyepetes Amataria.
Chesias Spartiaria. On Broom at night in October, Railway Banks between Willington Station and North Stafford Junction.
Odezia Chœrophyllaria. Mowing Grass, etc.

TORTRICINA.

Halias Prasiana.   Repton Shrubs, on Nut trees.
Tortrix Pyrastrana.
,,   *Xylosteana.*
,,   *Rosana.*
,,   *Ribeana.*
,,   *Corylana.*
,,   *Spectrana.*
,,   *Icterana.*
,,   *Viridana.*
,,   *Ministrana.*
,,   *Adjunctana.*
Peronea Favillaceana
,,   *Schalleriana.*
,,   *Comparana.*   Amongst Raspberries.
,,   *Abildgaardana.*   In variety.
,,   *Tristana ?*
Teras Effractana.   Potlock Covert.
,,   *Cuadana.*   The Henhurst, near Burton-on-Trent.
Dictyopterix Contaminana.
,,   *Læflingiana.*
,,   *Holmiana.*
,,   *Bergmanniana.*
,,   *Forskaleana.*
Argyrotoza Conwayana.
Ptycholoma Lecheana.   Repton Shrubs.
Penthina Pruniana.
,,   *Cynosbana.*
Spilonota Ocellana.
,,   *Dealbana ?*
,,   *Neglectana.*
,,   *Roborana*

Pardia Tripunctana.
Natocelia Udmanniana.
Sericoris Lacunana.
Mixodia Ratzburgiana.
Phtheocroa Rugosana. On Nettles.
Cnephasia Musculana.
Sciaphila Nubilana. Flying gregariously round shoots of Hawthorn.
Sciaphila Virgaureana.
„ *Hybridana.*
„ *Alternana.*
Sphaleroptera Ictericana. In meadows. Larvæ in flowers of Buttercup.
Capua Ochraceana. Two in Repton Shrubs.
Bactra Lanceolana.
Phoxopteryx Biarcuana. Amongst Wild Raspberries. Potlock Covert.
Phoxopteryx Lundana.
„ *Diminutana.*
„ *Metterbacheriana.*
Grapholita Paykulliana.
„ *Campoliliana.* On Willows.
„ *Trimaculana.*
„ *Penkleriana.*
„ *Obtusana.*
„ *Nœvana.*
Phlœodes Immundana.
Batodes Augustiorana. The only Lepidopterous Larvæ I have found on the Yew. The larvæ feeding through the winter on Yew leaves, which they have previously drawn together.

Pœdisca Corticana.
,, *Solandriana,* and varieties.
,, *Sordidana.* October and November. Near Nut bushes.
Ephippiphora Dissimilana.
,, *Nigricostana.* Findern Covert, etc. Occasionally abundant.
Ephippiphora Brunnichiana.
,, *Scutulana.*
Olindia Ulmana.
Semasia Woeberana. Near Repton. (P. B. M.)
Coccyx Argyrana. Findern.
,, *Hercyniana.*
Retinia Buoliana. Bretby Park.
,, *Pinivorana.* Between Willington and Etwall.
Endopisa Nebritana.
Stigmonota Perlepidana. On the young shoots of old Willow stumps, April, Repton.
Stigmonota Redimitana.
Dicrorampha Sequana? On white Clover. Railway Banks, etc.
,, *Senectana.*
,, *Plumbagana.* On Tansey.
Pyrodes Rheediana. On young shoots of Hawthorn, in sunshine.
Catoptria Ulicetana. On Gorse, when in flower.
,, *Hypericana.*
,. *Hohenwarthiana.*
Trycheris Mediana. On umbelliferous plants, by day.
Choreutes Scintillulana. Borders of lake in Bretby Park.
Simœthis Fabriciana.

Eupœcilia Maculosana. Repton Shrubs. Amongst Wild Hyacinths, in bloom.
Eupœcilia Angustana. On clover: railway banks.
„ *Roseana.* "The Oaks," near Burton-on-Trent.
Xanthosetia Hamana.
Cochylis Stramineana?
Aphelia Pratana.
Tortricodes Hyemana. Findern; Repton Shrubs; flying by day in groups. March.

TINEINA.

Exapate Gelatella. Repton Bridge Road in November.
Chimabacche Phyganella. Repton Bridge Road, and in the Potlock Covert, in October.
Chimabacche Fagella. On trunks of trees in March.
Semioscopis Steinkellneriana. On Hawthorn in April.
Tinea Oehlmanniella.
„ *Masculella.*
„ *Rusticella.*
„ *Arcella.*
„ *Cloacella.*
„ *Fuscipunctella.*
„ *Pellionella.*
„ *Semifulvella.*
„ *Biselliella.*
„ *Ganomella.*
„ *Comptella.*
„ *Cœsiella.*
Ochsenheimeria Birdella.
„ *Vacculella.*

Micropteryx Calthella. In the flowers of "King cups."
,, *Seppella.* Repton Shrubs.
,, *Allionella.* One in Repton Shrubs, on Honeysuckle.
Micropteryx Rubrifasciella. Repton Shrubs.
,, *Subpurpurella.*
,, *Fastuosella.* Near Repton. (P. B. M.)
Nematopogon Swammerdammellus.
,, *Sericinellus.*
,, *Schwarziellus.*
Adela Fibulella. On the flowers of Blue Veronica.
,, *Rufimitrella.* On Cardamine.
,, *De Geerella.*
,, *Viridella.* Repton Shrubs.
Plutella Cruciferarum.
,, *Sequella.* Near Repton. (P. B. M.)
,, *Vittella.*
,, *Fissella.*
,, *Costella.*
,, *Harpella.*
Ypsolophus Durdhamellus. Willington.
Anarsia Spartiella. On Furze in railway cuttings.
Œcophora Sulphurella.
,, *Oppositella.*
,, *Quadripunctella.*
,, *Flavimaculella.*
,, *Pseudospretella.*
,, *Lacteella.*
,, *Fuscocuprea.*
,, *Curtisella.*
Hyponomeuta Cognatellus.
,, *Padellus.*

Orthotælia Sparganiella.   Near Repton.   (P. B. M.)
Depressaria Costosa.
",,   *Liturella.*
",,   *Ulicetella.*
",,   *Assimilella.*
",,   *Arenella.*
",,   *Subpropinquella.*
",,   *Alstrœmeriana.*
",,   *Hypericella.*
",,   *Angelicella.*
",,   *Applana.*
",,   *Chærophylli.*
",,   *Ultimella.*
",,   *Heracleana.*
Carcina Fagana.
Gelechia Populella.
",,   *Cinerella.*
",,   *Gallinella.*
",,   *Terrella.*
",,   *Proximella.*
",,   *Mouffetella.*
",,   *Conscriptella.*
",,   *Ligulella.*   Bretby Lake.
",,   *Affinis.*
",,   *Tenebrella.*
",,   *Luculella.*   On trunks of trees in Repton Shrubs.
",,   *Lucidella.*   Bretby Park.
",,   *Lutulentella.*   On heads of larger Cuckoo Weed, in September.
",,   *Vulgella.*
Röslerstammia Fuscoviridella.

LEPIDOPTERA.

Gliphepteryx Variella.   Abundant on Wild Raspberries.
Æchmia Thrasonella ?   On grass in Repton Shrubs.
  ,,   *Fischeriella.*   On grass in Repton Shrubs.
  ,,   *Sericiella.*   On grass in Repton Shrubs.
Argyresthia Nitidella.
  ,,   *Albistria.*
  ,,   *Curvella.*   On flowers of Mountain Ash.
  ,,   *Goedartella.*   On Alder.
  ,,   *Brockeella.*
Ocnerostoma Piniariella.
Coleophora Spissicornis.
  ,,   *Alcyonipennella.*
  ,,   *Annatipenella.*   In cases on Hawthorn.
  ,,   *Onosmella*
  ,,   *Discordella.*
  ,,   *Alticolella.*
  ,,   *Cæspititiella.*
  ,,   *Laricella.*
  ,,   *Nigricella.*   On Elm.
  ,,   *Fuscedinella.*
  ,,   *Binderella.*
  ,,   *Lusciniæpennella.*
  ,,   *Lutarea.*
Zelleria Insignipenella.
Gracilaria Franckella.
  ,,   *Stigmatella.*
  ,,   *Syringella.*
  ,,   *Tringipennella.*
  ,,   *Auroguttella.*
Ornix Guttea.   Findern.
  ,,   *Meleagripennella ?*
  ,,   *Anglicella.*

Cosmopteryx Præangusta. One near Newton.
Laverna Ochraceella.
,, *Propinquella.*
,, *Epilobiella.*
,, *Vinolentella.* Near Repton. (P. B. M.)
Elachista Testaceella.
,, *Atra.*
,, *Locupletella.* Foremark.
,, *Festaliella.*
,, *Albifrontella.*
,, *Cinereopunctilla.*
,, *Pulchella.*
,, *Cerusella.*
,, *Ruficinerea.*
,, *Cygnipennella.*
Phyllocnistis Suffusella.
Cemiostoma Spartifoliella.
,, *Laburnella.*
,, *Scitella.*
Bucculatrix Cidarella.
,, *Cratægi.*
,, *Boyerella.*
Neptecula Ruficapitella.
,, *Ignobilella.*
,, *Floslactella.*
,, *Aurella.* Near Repton. (P. B. M.)
,, *Splendidissimella.* Near Repton. (P. B. M.)
,, *Angulifasciella.* Near Repton. (P. B. M.)
Tischeria Complanella. Repton Shrubs.
Lithocolletis Pomifoliella.
,, *Faginella.*

Lithocolletis Salicicolella.
,, *Spinolella.*
,, *Quercifoliella.*
,, *Corylifoliella.*
,, *Viminiella.*
,, *Cramerella.*
,, *Sylvella.*
,, *Tristrigella.*

www.ingramcontent.com/pod-product-compliance
Lightning Source LLC
Chambersburg PA
CBHW032128160426
43197CB00008B/565